AF866248

A Field Guide to the Marine Life of Coastal Reefs and Rock Pools

Phil Colman and Peter Mitchell

A Field Guide to the Marine Life of Coastal Reefs and Rock Pools

First published by Groundtruth Consulting as *Exploring Tidal Waters on Australia's Temperate Coast*

This edition published by:
David M. Welch
PO Box 503, Coolalinga, Northern Territory, Australia, 0839
www.aboriginalculture.com.au
www.davidmwelch.com.au

National Library of Australia Cataloguing-in-Publication entry

Authors:	Phil Colman and Peter Mitchell
Title:	A Field Guide to the Marine Life of Coastal Reefs and Rock Pools
Edition:	2nd ed.
ISBN:	978-0-6455122-7-4 (pbk.)
Notes:	Includes bibliographical references, glossary and index
Subjects:	Seashore ecology – Australia
	Geomorphology – Australia
	Coasts – Australia

Dewey Number: 577.6990994

Designed and typeset in 11.5pt Times Roman by Bruce Welch.

Printed in China on 115 gsm Lumi Silk artpaper, section sewn.

Front cover photographs: Top Row - *Octopus tetricus*, D Thomas; *Glaucus atlanticus*. **Bottom Row -** *Ceratosoma amoenum*, D Thomas; *Doriprismatica atromarginata*, D Thomas.

Back cover photo: Sea urchin test and crab claw amongst coarse shell debris.

All photos and drawings are the authors' except where otherwise credited.

Acknowledgments

Over the years we have become indebted to a great many people who have discussed our ideas for this book and sometimes we never quite know where a particular thought originated. We thank you all.

As the dream came closer to reality a number of people made substantive contributions and we would like to thank them individually:

Ruth Colman, Wendy Orchard and Fiona van Dort for their deft editing.

Cathy Hemery and staff at the Coastal Environment Centre for much needed reader's feedback.

Christine Pedersen for a very enthusiastic lay-person's critique that drew our attention to difficult words.

Peter Donkers, Jack Hannan, George Evatt, Dave Thomas, Cherry and Rowan Morrison, Peter Miller, Vera O'Donovan, and Tom Brougham for photographs we were unable to get.

Dr Mark Norman for identification of octopus and cuttlefish photographs and Dr Bill Rudman for identification of sea slugs.

Alan Miller, Ricki Coughlan, Jo Adendorff, Anthony Davis, and numerous others for conversations that none of us ever imagined might one day find their way into print.

John Illingsworth and Paul Cronk for filming and mental challenges.

And to Bruce Welch for efficiently guiding the publishing end of things.

We hope you all enjoy it.

Phil Colman, Peter Mitchell

A quick first word

This is a new edition of a book we privately published in 2011. That was successful in that we managed to sell all copies and even made a small profit on the exercise, although that was never our objective. All we ever wanted to do was inspire shore platform visitors to look at the wonders of the inter-tidal zone more closely and to be guided by Rudyard Kipling's Six Honest Serving Men:

> I KEEP six honest serving-men
> (They taught me all I knew);
> Their names are What and Why and When
> And How and Where and Who …

Phil had been taking people over the Long Reef Shore platform (Sydney) for the best part of half a century and the original purpose of the book was to provide plain English answers to questions he met time and again. The idea grew and Phil wrote short accounts on many topics then it became a joint effort when Peter turned up with geological knowledge and suggested that it should all be run together and illustrated.

Just like the organisms we talk about the book grew, drafts came and went and it got bigger. We drew most of our examples from Long Reef because we knew it well and then we slowly found ourselves talking about places in Victoria and New Zealand where the stories were very similar.

A South African friend wrote to say that he was quite happy to use our book with his students on rocky shores and we realized that Long Reef was just a convenient example of most of the processes and relationships that can be seen almost anywhere in the temperate inter-tidal world. Patterns of species distribution, life strategies, dispersal methods, modes of communication, and species survival stories are much the same in the Americas, Africa, Asia, Australia, New Zealand, and even in the Mediterranean. The individual species are different but the same families occupy the same niches in all those places. Littorinids are always 'up there', barnacles cluster at the same levels of the tide and waves, and similar looking kelp lives just below the low tide level.

This version of the book has grown some more (and we have corrected a few errors in the first version), and we are very aware that there are another 10 or 20 topics we could open but we hope that there is enough here to send you off on an enthusiastic search to discover something new for yourself.

Phil Colman, Peter Mitchell

Contents

One of many beautiful plates of natural history specimens by Ernst Haeckel (1834-1919). Haeckel's assistant and student, the Russian scientist and New Guinea explorer Nikolai Nikolaevich de Miklouho-Maclay established Australia's first marine biological research station at Watsons Bay in 1879.

Top left Juvenile *Floresca* sp. Centre Adult lion's mane jellyfish *Cyanea annasethe*. Haeckel named this North Sea species after his first wife of 18 months who died shortly before he described it. Bottom *Cyanea annasethe* dissected. Right Adult compass jellyfish *Chrysaora hysocella*, Europe and the Mediterranean. This genus is also found around southern Australian. From: Kunstformen der Natur (1904), plate 8: Discomedusae. Wikimedia Commons.

Setting the scene

One third of the New South Wales coast is rocky cliffs and shore platforms, but most people ignore these and go to the beach. We challenge you to visit a shore platform during low tide and discover some of the magic. A great place to start is our favourite shore platform at Long Reef, but anywhere will do as similar scenes and creatures can be found all along our temperate coast, and in many other parts of the world.

Ardea novaehollandiae standing in the cunjevoi and *Caulerpa filiformis*, at low tide with an eye on dinner.

Viewed on a warm summer afternoon, with the ebbing tide and a white-faced heron *Ardea novaehollandiae* placidly browsing in a rock pool, the shore platform is really a place of great activity. There are constant quiet battles for supremacy between dissimilar organisms. Species joust for space, and there sometimes seems to be a multi-sided skirmish when two organisms appear to collude to exclude a third. But if that is so, the 'friendship' shown between the allies is soon lost in the next bout.

During one visit the platform may be covered in oysters. Six months later the oysters have disappeared, and limpets have taken their place. Several months on, and algae dominate, then barnacles. But it is not necessarily the oysters that return next. The battles are rarely so decisive, and oysters might or might not get another chance before they are again subjected to total ostracism or accept a minor role.

What is happening? To a certain extent it is the old 'eat-or-be-eaten' syndrome. Oysters may be there because there was a good recruitment of juveniles aided by pheromones given out by the adults already present sending the message that 'this is a good place to be'. But carnivorous mulberry whelks (aka black oyster-borers) *Morula marginalba* that can be found in all the shore platform environments, - oh dear, there's our first problem. These gastropods actually belong in the murex family not the whelk family so we will set a new standard and simply call them mulberry snails. Somehow the mulberry snails learn of this bonanza, and arrive to feed on oysters. As dead oyster shells are washed from the rock they make space for algal spores. But algae are ephemeral and die off with a weather change, perhaps making room for barnacles to expand.

A lone *Morula marginalba* nestles amongst its barnacle prey *Tesseropora rosea.*

However, mulberry snails also have a liking for barnacles, and the scene changes again as the rocks are cleared of barnacles. Mother Nature abhors a vacuum, and limpets move in keeping algae at bay by eating the juveniles. Mulberry snails like limpets too, but it could be that oystercatchers, birds that don't eat oysters (but more of that later), decimate the limpets.

So life goes on, changing all the time as one organism fights for supremacy only to be deposed by another, and another, and another, and …

Shore platforms

The physical shape and extent of shore platforms depends on the local geology and the power of eroding waves. Before using Long Reef as a more detailed example, consider the following cases where rock type and rock structure essentially control the topography of the shore platform.

Platforms below hard sandstone cliffs along much of the Sydney coast plunge steeply into the sea or may be buried by huge boulders fallen from the cliff. Although favoured by rock fishermen these places are often hard to access and dangerous to visit. They are best admired from a distance.

Cliff and shore platform on quartz sandstone at Little Bay, Sydney.

Where softer sandstone and shale beds lie horizontally very wide platforms are cut at about the mid-tide level. Typically they rise gently to the base of the cliffs along an abrasion ramp and they may have a slightly raised rampart at the low-tide edge where the rocks remain saturated and weather a little more slowly. Gutters and pools are present but large flats that are eroded as the rocks slake through repeated cycles of wetting and drying may be almost featureless.

A horizontal shore platform on purple shale at Turimetta Head, Sydney. The large sandstone blocks have fallen from the cliff.

Shore platforms on tough and deformed (folded and faulted) sedimentary and metamorphic rocks often form long rock ridges separated by numerous gutters. The structure of the rock and the different hardness of individual beds controls the shape A good example is the platform eroded into chert, conglomerate and slate of Cambrian age at Barlings Beach north of Moruya, New South Wales (NSW).

Rock structure controls details of the topography of the shore platform eroded into hard vertical strata at Barlings Beach, NSW.

Planes of weakness in the rocks (joint planes and bedding surfaces) always influence the shape of the platforms. This is particularly evident in igneous rocks such as basalt and granite where the joint planes often have regular geometry.

The spray zone (supralittoral) extends higher than usual on this steep shore platform formed on columnar jointed volcanic rock at Kiama, NSW.

Good examples can be seen on columnar volcanic rock at Kiama. Platforms on granite often have very large originally cubic blocks that are rounded off by more rapid weathering from three directions at the corners. At Nundera Point on the south coast of NSW a dark coarse grained plutonic rock called essexite weathers into rounded forms much like granite.

Cubic joint patterns in coarse grained igneous rock shape this stepped and gently sloping platform with large boulders at Nundera Point, NSW. The absence of lichen on the boulders indicates that they are moved by storm waves.

In weak rocks and with powerful waves it is possible to have a cliffed coast with almost no shore platforms because the rocks weather so rapidly that they are reduced to sediment and washed away. The narrow platforms and beaches and the cliffs and tall rock stacks along the well-known Port Campbell coast in western Victoria are good examples. In the extreme case of very soft rocks and low wave energy the shore platforms can be composed of soft rock, sand and mud interspersed with occasional rock bars. The muddy shores in Auckland Harbour that are rich in burrowing organisms illustrate this.

Elsewhere in New Zealand such as the wide platforms and boulder beaches near Wellington on the North Island, and in cemented mudstone and limestone at Kaikoura on the east coast of the South Island the coastal landscape has been complicated by recent earth movements. These platforms and beaches are stepped with each step reflecting a vertical movement of the land. The cold water on these shores is rich in nutrients and the shore platforms have very high diversity of organisms many of which will be familiar to an Australian visitor. Turakirae Head near Wellington (NZ) has three raised shore platforms and the 2016 Christchurch earthquake lifted the platform at Kaikoura by 5.5m.

Ribs of slightly harder rock are colonized by oysters on the muddy shore platform at Big Shoal Bay in Auckland Harbour, New Zealand. Photo by Cherry and Rowan Morrison.

Jargon

In writing a book like this we need to use some special words. Scientific names are important, although we have done our best to eliminate jargon. But some terms simply don't have common alternatives. The scientific names we have used are all up to date but they do change and there is a risk that what we write today will be out of date tomorrow. Where a species name has recently changed or another name was widely used we have placed the old name(s) in brackets as they may be familiar to you or can be found in older books.

What is a Species?

Species are the basic units of biological classification, each being a single kind of plant or animal that is capable of producing viable offspring. Other organisms that are very similar but that are reproductively isolated from each other belong in the same genus. For example the commercial blacklip abalone in New South Wales is *Haliotis rubra.* Paua, its colourful New Zealand cousin, is *Haliotis iris* and the Californian black abalone is *Haliotis cracherodii.* Under our binomial system of classification all organisms have two names, firstly the genus and secondly the species. In writing, the genus name begins with a capital, then the species name in lower case, all in italics or underlined. The generic name is spelled in full on the first occasion but can be represented by its first letter from then on (*Haliotis rubra, H. cracherodii,* and *H. iris*). To be strictly correct the name and date of the author(s) who first described the species should be added. For example, the proper name of the violet snail with a float of frothy bubbles that washes onto local beaches after drifting in as part of the plankton is *Janthina janthina* (Linnaeus 1758). Yes, Carl Linnaeus described it 12 years before James Cook sailed the *Endeavour* into Botany Bay!

Left: *Haliotis rubra* from Australia.
Right: *H. iris* from New Zealand.

The plural of genus is genera; closely related genera belong to a family. Related families belong to a Superfamily, then an Order, then a Class and finally a Phylum. So abalone, garden snails and periwinkles, all having one shell in which the animal lives, belong to the Class Gastropoda. Oysters, giant clams and freshwater mussels are shellfish too, but they all have two shells, so they belong to the Class Bivalvia. Octopus and squid also have certain things in common so they belong to the Class Cephalopoda. All three groups, related as they are, belong to the phylum Mollusca in the kingdom Animalia (Metazoa). Bluebottles, coral polyps, and jellyfish likewise have certain things in common (stinging cells or nematocysts are peculiar to them) and belong to the phylum Cnidaria (pronounced with a silent 'C').

Janthina janthina, note the colour difference (counter-shading) on the top and bottom of the shell.

Don't be put off by scientific names. They are not as formidable as they appear. You already use some in everyday language, for example, *Eucalyptus*, *Acacia, Jacaranda* and *Bacterium.* Take it a step further, *Eucalyptus saligna, E. robusta*, *E. regnans*. Note that the genus, or generic name, remains the same in each case, so that half the battle is won when you get that far.

Scientific names may be a bit daunting but they are better than common names that sometimes are only used locally and are certainly not standardised. For example, a common bivalve used as fishing bait along our NSW shores is known as a pipi (also pippi). This just happens to be a Maori word that applies to three different bivalve genera in New Zealand. That raises the intriguing question of why was that name applied in New South Wales? In South Australia the same shellfish is called a Goolwa cockle, while in Queensland it is sometimes known by an Aboriginal name, ugari (also eugari or yugari). But any scientist in Sydney, Tierra del Fuego, or London can avoid this confusion because everywhere in the world it is *Donax deltoides* as there is only one such species.

Try to get used to scientific names. Four-year-olds have no trouble with *Tyrannosaurus rex,* so give it a go.

Donax deltoides live in the sand in the wave zone and are not normally seen on the surface. The photographer had to be quick to catch them before they upended themselves and dug for home.

How many marine species are there? Well, we don't know and every source you check has wildly different estimates. Figures for the molluscs for example range from 50,000 to more than 200,000. One thing for sure though, our temperate shore platforms are some of the most biodiverse places on Earth and yet we only have reliable names and descriptions for about one third of the organisms living there.

Classification of living things

The hierarchy is: Kingdom, Phylum, Class, Order, Family, Genus, species, and sometimes subspecies or variety.

Try it with humans: Kingdom: Animalia or Metazoa, Phylum: Chordata, Sub-phylum: Vertebrata, Class: Mammalia, Order: Primates, Family: Hominidae, Genus: *Homo*, species: *sapiens*.

sapiens being a Latin word meaning 'being wise'. We sometimes wonder.

An Aboriginal past

Humans are naturally attracted to the water's edge, and where better for indigenous people to go for a feed than to a wide coastal shore platform with a diversity of life forms that are easily accessible? Right around Australia Aboriginal people used accessible shore platforms as important food resources collecting shellfish and catching fish.

Close to resource rich shore platforms evidence of occupation by Aboriginal people is usually found as shell middens composed of the remains of meals and charcoal from fires. Coastal middens near Sydney contain at least 35 species of fish, fragments of crustacea, and large numbers of shellfish. The shells include shore platform species that we immediately recognise as food such as blacklip abalone *Haliotis rubra,* the Sydney rock oyster *Saccostrea glomerata* (*S. commercialis, Ostrea commercialis*), and the cartrut shell *Dicathais orbita (Thais orbita).* Shell middens also contain species that we do not think of as food. Examples are: the hairy mussel *Trichomya hirsuta*, the black nerite *Nerita atramentosa*, the variegated limpet *Cellana tramoserica*, and a surf barnacle *Austrobalanus imperator.*

Shells in the midden do not represent the whole diet but only that part of the diet with durable remains. No certain trace of cunjevoi *Pyura praeputialis* has been recorded but we would certainly expect it to have been eaten. We have little knowledge of the use of seaweed, or coastal plants, and the remains of smaller fish and crustacea must be under-represented in comparison to larger shellfish. What we can say however is that the food items were drawn from all parts of the shore platform, even into the sub-tidal water where the collection of abalone and southern rock lobster *Jasus edwardsii* would have required shallow diving.

Middens also contain stone flakes, burnt hearth-stones and artefacts. These include bone points that were attached to four pronged fishing spears, unbarbed fish hooks made from the Sydney turban shell *Turbo torquatus,* and small stone files that were used to shape them. Items made of less durable material such as wood, bark, fibre or hair are rarely preserved. Aboriginal burials occur in larger middens and scattered human bones have been recorded near the surface of many middens near Sydney. These are believed to be the remains of smallpox victims who died in 1789-1790 and were never buried.

Because sea level was so much lower during the Ice Age most coastal middens are less than 2,000 years old and the oldest with shells are only 4,000 years old.

At the time of European settlement shell-middens were common along the Sydney coast, but many were excavated and burnt for lime, or destroyed by urban development, sand mining, and coastal erosion. All of those that remain should be respected as Aboriginal sites. Middens are protected by law, but the law will not save those that are closer to the water and threatened by rising sea levels driven by climate change.

Aboriginal midden composed of shell, ash, and charcoal at Kurnell, NSW. The large rock is sea-rafted pumice and above that are seal rib-bones. In 1770 James Cook encountered Gweagal Aboriginal people on this campsite. The three highest layers contain no Aboriginal material as they were all deposited after about 1840.

The water environment

Shore platforms lie at the boundary between three very different worlds. Below the lowest tide we have oceanic conditions where fish reign, above the tide are terrestrial systems where plants reign, and above both is the aerial environment of the birds.

Long Reef is flushed twice daily by tides and less frequently by rain. Organisms living on the platform can be baked by the sun, soaked in warm brine, battered by powerful waves, abraded by sand and gravel, drenched in rain, and trampled by visitors all in a single day. It is a tough place to live but it also contains numerous microenvironments where many different organisms can survive. A huge diversity of life forms can be found and a platform like Long Reef may support 10,000 visible species, perhaps making it one of the most biodiverse places on earth.

The ups and downs of sea level

When the level of the sea changes the intertidal zone of the shore platform shifts, so places that are normally at the dry end or the wet end of the platform are subject to different conditions.

Sea level varies through many time cycles, the shortest being seconds between the peaks and troughs of the waves, over minutes with rare tsunami waves, twice daily through the tidal range, and over years linked to El Niño weather patterns.

Even bigger sea level changes were driven by the Ice Ages as sea level fell around the globe as polar ice caps expanded. On the east Australian coast geological evidence shows that sea level during the last 11 interglacial periods (warm global climates) was never higher than 6m above the present level. Between 125,000 and 140,000 years ago it was three to five metres higher and 18,000 to 20,000 years ago at the peak of the last glacial period (Ice Age) it fell 120 to 130 metres below the present level. This placed the Sydney coastline 10 to 12 kilometres off the present shore and Long Reef was just another rocky hill separating the valleys of Narrabeen and Dee Why Creeks. As the Earth warmed, sea level rose isolating Papua New Guinea and Tasmania. It then exceeded the present sea level by about 2m between 8,000 and 2,500 years ago. This small shift is not easy to recognize today as much of the evidence has vanished.

The future however will not necessarily be a repeat of the past. Tide records show that average sea level at Sydney has risen more than 100mm in the last century, apparently as a result of global warming. Computer modelling of climate change indicates that we can expect another 400mm rise by 2050. The models are less reliable to 2100 when sea level may be close to one metre higher than today. Coastal councils in NSW are using these figures as planning guidelines. Even greater increases of six to eight metres or more by 2100 have been predicted and although very difficult to accept, they are not impossible.

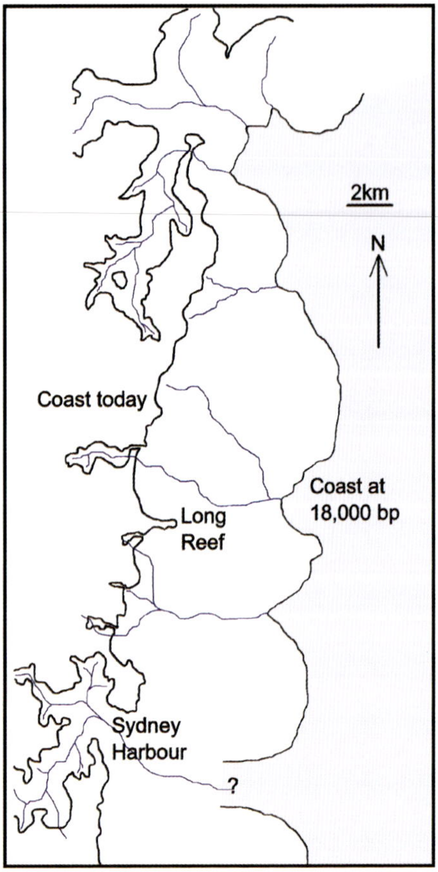

Outline of the Sydney coast during the last Ice Age 18,000 years ago. A map like this can be drawn for most of the world's coastline.

Waves

Waves breaking at the coast carry the energy of local winds and distant storms. On the east coast the underlying pattern of swell waves is normally created by storms in the Southern Ocean. Waves travel in great circle routes around the world unaffected by the Coriolis 'force' and they only change direction when they 'feel bottom' near the shore. Then they become steeper and higher and curve (refract) to fit the coastline. Long Reef experiences waves from every offshore direction because it is such a prominent feature. Average deep-water wave height is 1.6m, common storm waves reach 4.5m and very big storms such as those of 1974 generate 8.5m waves that can cause a lot of coastal erosion.

Tides and currents

Tides are a natural rhythm of the ocean driven by a combination of the gravitational pull of the Moon and Sun on the waters of the Earth. Gravity pulls a bulge of ocean water around the equator. This combines with a centrifugal bulge of water pushed out by the spinning Earth and the normal result is two tidal cycles every day. When the Sun and Moon are in alignment (full Moon or new Moon) their combined gravity pulls in the same direction and we have the maximum tidal range called spring tides (twice each lunar month). When the

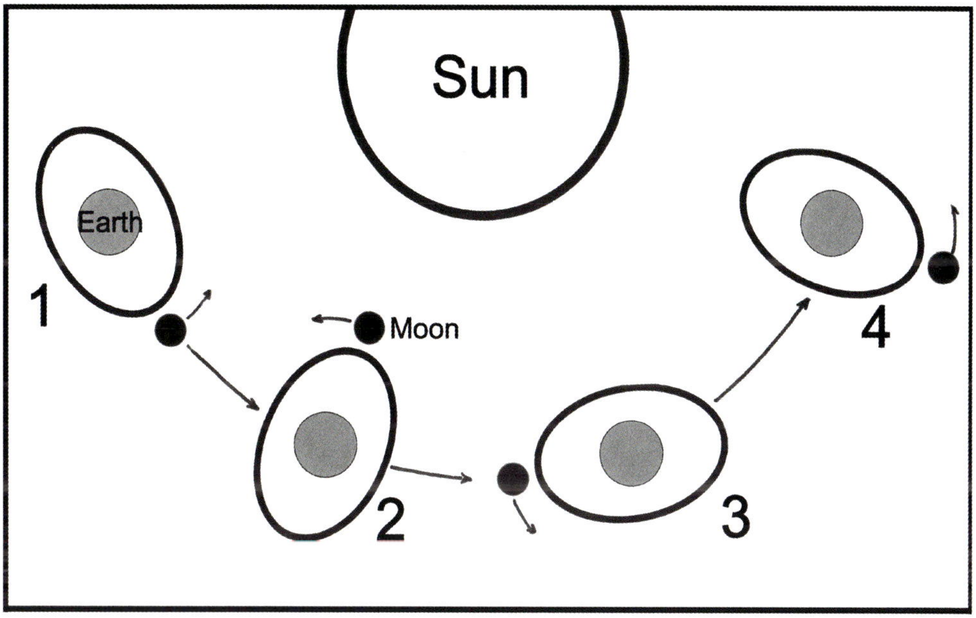

The global tidal bulge formed by the alignment of the Earth, Moon and Sun. Neap tides occur at positions 1 and 3, spring tides occur at 2 and 4.

Sun and Moon are at 90^0 to one another (first and third quarters) they pull in different directions, the tidal range is reduced and we have neap tides. During neap tides organisms living high on the shore platform may be exposed to the sun and air for quite long periods. When tides are at their highest and lowest in a year they are called king tides. These occur in June and December when the Moon is a little closer to the Earth.

Tides are not uniform in time or space as they vary with the shape of the ocean basins and the movement of water masses by the Coriolis 'force'. Tides have a 25-hour Lunar cycle therefore the time of high and low tide shifts forward about one hour every day. The entire east coast of Australia experiences tidal peaks at almost the same time and the tidal range varies from almost zero metres to just over two metres measured against the Australian Height Datum.

Tidal rhythm can play a big role in the activities of organisms. Fishermen know this. Different species will only bite on the high tide, or near the end of the ebb tide. Intertidal animals are also influenced. One may seek for hours in a large pool for a particular animal, as the tide goes out, with no success. Then the tide turns and that formerly elusive animal starts to appear everywhere. What is happening is that that animal, possibly considered a tasty morsal by many predators, emerges to carry out life's processes be it feeding, mating or whatever, before the higher tide allows predator access, when it again goes into hiding.

The oceans of the world are in constant circulation through a network of surface and deep-water currents driven by prevailing winds and water density. East coast currents shift warm water toward the poles, and west coast currents move cold water north. The net effect is a redistribution of stored solar energy and nutrients, moderation of coastal climates, concentration or dispersal of fish stocks, and the movement of marine species from place to place. Cold deep water, which contains the greatest quantity of nutrients gets into the act through the process of upwelling with the classic example being the growth and collapse of the Californian sardine industry and the Peruvian anchovy fishery that replaced it. There was of course a triple-whammy in that story which involved human greed, uncontrolled harvest, inappropriate use of the product, and a fishing technology that was immediately available to move from the US to South America when the northern stocks collapsed.

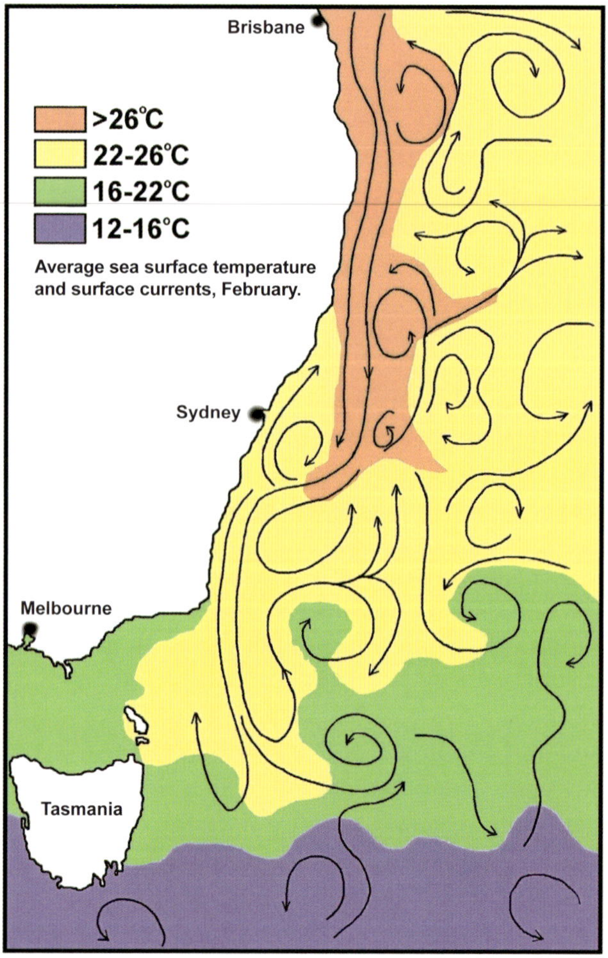

Average February water temperatures and surface flow in the EAC as it was in 2000. By 2020 both the yellow and red zones had extended 300km further south.

Australia is bathed by the East Australian Current (EAC), the Flinders Current across the Great Australian Bight, and the cold West Australian Current which is not a typical west coast system as it interacts with the warm Leeuwin Current which flows south tending to suppress upwelling. The east and west coast currents are marine highways for migrating whales, especially humpbacks.

In recent decades we have learned a great deal about the EAC, just at the time when its physical nature is changing as the climate warms. The EAC begins where the South Equatorial Current crosses the Coral Sea and splits against the Great Barrier Reef and the Queensland coast. It flows south as a broad, warm, low nutrient, surface current 100 km wide and 500-1,000 m deep. It is strongest in summer and moves at about 3 km/hour.

In southern NSW the EAC breaks into multiple warm or cold core eddies and then dissipates into the Tasman Sea. But it is changing. Warm water now frequently passes Bass Strait. The cause is global warming and the consequences are a redistribution of marine species, and general warming of the east coast climate. Three jellyfish *Chrysaora sp., Catostylus mosaicus* and *Physalia physalis* all common around Sydney are now frequently seen in Tasmania, and a tropical manta ray and turtles were sighted there in 2016.

By far the biggest effect has been the demise of the formerly luxuriant forests of giant kelp *Macrocystis pyrifera.* Giant kelp fronds can grow to 35m, and the forests were the basis of an entire food web. As they declined, long-spine urchins *Centrostephanus rodgersii* moved in and grazed kelp regeneration to the point that extensive 'barren grounds' were created. Kelp forest decline is not limited to Tasmania. It is of concern in Japan, Alaska, California and Norway where they also suffer a decline in lobster and pelagic fishery production.

Currents move all sorts of things very long distances. Two sub-fossil eggs of the elephant bird *Aepyornis maximus* were found in sand dunes in Western Australia in 1930 and 1992 and could only have come from Madagascar. Bull kelp *Durvillaea antarctica* was recently found on the Antarctic coast having travelled thousands of kilometres from Kerguelen Island and crossing the Antarctic Circumpolar Current which used to be considered as a barrier to species dispersal. And it carried invertebrate passengers!

Change is always with us.

Long Reef

The Long Reef shore platform 18 kilometres northeast of Sydney is one of the largest platforms in NSW. We will use it to illustrate common features that you may find on any platform.

Geologically it is formed on sub-horizontal beds of conglomerate, sandstone, and fine-grained shale. These rocks are known as the Bald Hill Claystone, part of the Narrabeen Group that is about 240 million years old (Triassic). First described in 1878 by Australia's pioneer geologist the Reverend W. B. Clarke as 'chocolate shale' (referring to the colour), this rock unit extends from Wollongong to Patonga, although it is not visible as a continuous body. At Long Reef it is only 18m thick and the lowest part is a dull green conglomerate that merges upward into mottled purple-brown claystone. The sequence is exposed in gentle folds and has been broken by several small faults and an igneous dyke (a vertical sheet of volcanic rock). At the eastern point of Long Reef a small outcrop of quartz sandstone known as the Bulga Sandstone forms a prominent ridge at the waters edge and an island at high tide.

Fossils: traces of the past

True fossils are rare in the rocks at Long Reef, but trace fossils, tracks, trails and burrows preserved in the rocks are reasonably common. Fossil amphibian bones have been found and a poorly preserved fossil crustacean about 250mm long was found in a fossil burrow called *Turimettichnus*. Yes it's strange, the burrow has been named as a trace fossil but not the animal! Fragmented plant fossils can be found on the bedding planes in the shale. These include a horsetail reed *Phyllotheca*, a shrub like seed fern *Dicroidium*, poorly defined root traces, and thin lenses of coal that are actually compacted logs. The fossils and the nature of the rocks suggest that the environment at the time of sediment deposition alternated between a riverbed and an estuary or a quiet coast. Both the leaf traces and the coal are sometimes associated with thin films of native copper or blue-green stains of copper salts that led optimistic prospectors to drive a tunnel into the southern face of the headland more than a century ago. There was no ore body.

In addition to biological trace fossils shallow water ripple marks are often preserved in the rock and you can find ancient mud cracks that have been filled with sand.

Turimettlchnus - filled fossil burrow of a crustacean found in the Triassic Bald Hill Claystone at Turrimetta Head and Long Reef. The curved lines on the side are casts of the claw marks made in the mud when the burrow was inhabited, twenty cent coin for scale.

Fossil ripple marks preserved in siltstone. These are common in sedimentary rocks along the Sydney coast that were originally formed in shallow water.

Fossil mud cracks preserved by an infill of sand, Turimetta Head, NSW.

Rock weathering and erosion processes

Rocks on the shore platform are susceptible to wetting and drying cycles that cause salt crystallization and surface slaking. With twice-daily tides and varying wave and weather conditions higher parts of the platform are subject to about 300 daytime wetting and drying cycles every year (drying is less at night).

Waves also play a part in the break up of jointed rocks and the relocation of loose debris onto a beach or into deep water. Hydraulic pressure in joint cracks and bedding planes is sufficient to dislodge blocks of rock that then erode the surface as they are moved around. Medium sized boulders may move with every storm and deep scratches may be visible on the platform surface. Smaller particles such as gravel and sand may move with every wave. Toward the eastern end of Long Reef point a black sand beach can be seen where heavy minerals (mainly ilmenite - iron titanium oxide) have been concentrated by wave action.

Rapid weathering of the siltstone and mudstone on the Long Reef shore platform creates a concentration of heavy minerals in a black sand beach. The sandy slope on the right is only a thin sheet of sand blown up from the beach and burying the cliff.

Cliffs and the abrasion ramp

A retreating cliff lies at the back of the platform as waves undercut the soft claystone and easily weathered sandstone. Salt from ocean spray accelerates rock weathering on the cliff and the rock is weak enough to slump or fall. Wetting and drying cause the weathered rock to fret and even on dry sunny days fretted granules tumble down the scree aprons. Scrambling kids do their bit to assist. At the top of the Long Reef cliff weathering has concentrated iron oxides. When debris from this zone falls onto the shore platform it hardens into solid boulders that accumulate at the base of the cliff.

Abrasion ramp at the base of the cliff. Storm waves undercut the cliff and rapidly remove fallen debris, Long Reef NSW.

At the high tide limit the surface of the platform has a gentle slope seaward across which waves roll boulders and sand that plane the surface smooth. This is called an abrasion ramp. There is an irregular cycle of debris accumulation and subsequent removal by waves that eventually leave only the largest boulders in place. The platform on the southern side of Long Reef is narrower and has fewer boulders than the north side, and the cliff is steeper and less vegetated because southern storms sweep the platform clear.

The big flats

The biggest level areas (let's call them the big flats) are formed on fairly uniform reddish purple claystone (not quite shale). Shallow pools are left on this surface as the tide recedes. The smaller ones evaporate leaving a salt crust on hot days. Frequent wetting and drying cycles cause the rock to slake and fret creating the big flats by a process known as 'water-layer-weathering'.

Extensive flats planed off by 'water-layer-weathering' and well covered by *Hormosira banksii*, Long Reef NSW.

In winter and spring the flats are smothered in a compact pasture of olive green algae called Neptune's necklace *Hormosira banksii*, named after Sir Joseph Banks who collected the original specimens from Botany Bay in 1770. This brown algae (it actually looks green) is found from Albany in Western Australia all around the southern shores including Tasmania, up to Coffs Harbour, and across to New Zealand. Each frond looks like a row of beads about 12 to 15mm in diameter. On the surface of each bead you may see little yellow spots that mark the egg and sperm chambers. As you walk on them the vesicles split open with a popping noise and you may be surprised to find that they are not filled with air, but with water that keeps them alive whilst exposed at low tide. Seaweeds wilt on exposure and reduce their rates of photosynthesis. In a daytime low tide cycle, exposed *H. banksii* beads will partly collapse and may lose 40% of their fresh weight. They will then require an hour of rehydration before returning to normal rates of photosynthesis. Exposure to very high air temperatures (heat wave conditions) can be fatal.

The tangled mat that you see at low tide is nothing like the open canopy of waving fronds that you would see underwater when the tide is in. Part the mat carefully, look closely, and you will be amazed at the variety of other life forms living here.

Subtle variation in the rocks on the flat creates microhabitats that are important in increasing the diversity of organisms across the platform. Adjacent to Fishermans Beach the shale and lithic sandstone dips west at low angles and the platform has a series of steps with intervening ridges and gutters. These features give the low tide visitor a window into marine conditions that can be comfortably examined without swimming in the surf zone.

Beds of slightly different rocks slope (dip) gently to the right to form ledges and gutters that provide different living conditions, Long Reef NSW.

Gutters, gullies and crevices

Running across the flats and indenting the edge of the platform are numerous elongate gutters, gullies and crevices. These are formed along joint planes (a line of weakness perpendicular to the bedding) or at the margins of different rock types through which water courses back and forth with every wave, sometimes rapidly, sometimes gently. There are plenty of organisms clustered in these areas. Those species more reliant on water are near the bed of the gully. Those further away have greater tolerance to exposure. Absence of water can play a vital role in the survival of certain organisms. Even brief exposure to desiccation may mean death to delicate organisms while others flourish with no detrimental effects. In fact certain organisms need this exposure/flooding regime to exist. Parts of crevices may be exposed to high wave pressure, other parts can trap air pockets where even spiders and pseudo-scorpions can live. Some channels have abrasive sand or pebbles moving back and forth along their floor slowly widening them.

Prominent joint planes in sandstone control the edge of the shore platform at Norah Head, NSW.

Gullies running across the platform at right angles to the coast can provide a sample of all the zones across the tide range but even here there is subtlety in the distribution of species. It could be that exposure to full sun on one side of a small gully is enough to control the presence or absence of a certain species. Any caves or crevices of course will host those plants or animals intolerant of sun or partial shade. There are even a few species that are confined to the gullies, such as the cockle *Codakia rugifera* that appears to live in tough conditions in deeper sand that has low oxygen levels. The green coloured swift-footed crab *Leptograpsus variegatus* likes rocky faces and can even be seen feeding upside down on the roofs of small ledges.

A bivalve *Codakia rugifera* lives in the gullies in sand and gravel that is often anaerobic. This was one of 27 molluscs illustrated by Thomas Watling, a convict artist with the First Fleet.

Between the rock stacks (small islands at high tide) and the sand spit at the eastern point of the headland an igneous dyke of dolerite about one metre wide cuts through the rock as a vertical sheet. Surprisingly this has little influence on the shape of the platform, and the adjacent shale is weathered to almost the same level. The dyke does not take a straight line across the platform but is offset on the joints. It can be traced right across the platform and then is not seen again. It must be in deeper water in front of the north side of the shore platform. A few dolerite boulders are found in the boulder field east of Fishermans Beach and the same dyke occurs as a strongly weathered, iron cemented, dark rock mass on the beach near Florence Avenue.

An igneous dyke cuts across the horizontal sedimentary rocks. All the rocks are weathered and eroded to almost the same level on the Long Reef shore platform.

Rock pools

Larger, deeper pools in which marine life is abundant are found close to the platform edge where large joint blocks have been removed by storm waves. Typically these are square to rectangular and the depth can vary from one side to the other if the rock bedding is not horizontal.

A large square pool formed by the removal of a joint block, Long Reef, NSW.

'Cannon-ball' concretions formed in sandstone. When these are plucked out by waves a round pool can begin, Wollongong Harbour, NSW. Concretions quite often have fossils in their core.

A small round pool eroded into sandstone by swirling pebbles and shells, Long Reef, NSW.

In softer rocks, round pools with a bed of stones and shell may also be seen. These are maintained by rotating stones that slowly grind away the bottom of the pool. To start a pool there needs to be an initial depression in the rock. In time the pool may become deep enough for the stones to be immobile except during big storms and at this stage the pool will contain its highest diversity of organisms.

Very shallow pools that may evaporate during exposure are common across the big flats and elongate pools that often carry flowing water lie along the gullies.

The biodiverse world is richest when all factors combine. Marine biodiversity is probably at its maximum in pools just above low tide. Organisms can carry on living with less stress as these pools do not dry out and individuals have a chance to forage openly before the rising tide brings in the predators.

Nevertheless a saline rock pool can be a tough placc to live as it will go through drastic changes over a single tide. On an average summer day starting with low tide at say noon, some major changes take place in that pool. The day is hot. The pool is blocked from any interchange of water. So the temperature climbs, evaporation increases the salt content and decreases the dissolved oxygen. The surface temperature, which in a shallow pool means the whole pool, can exceed 35^{0}C around Sydney. The level of dissolved oxygen can fall from 20% to 8% and the pH of the water can change from 7.00 (neutral) to 10.0 (alkaline) in a few hours.

Things are not looking good. But more is to come. In the afternoon a summer storm looms in the west. Rain falls, the fresh water floods the pool, water temperature and salinity decrease and conditions shift from hot and hypersaline to cool and nearly fresh. Sometimes a hailstorm accompanies the weather change and this can impact directly on organisms on the rock surface. Water temperature can plummet 20^{0}C in as many minutes.

On warm sunny days water in pools containing seaweed may develop a patchy layer of foam on the surface. This is actually oxygen released by photosynthesis. Pools with large numbers of animals may accumulate so much ammonia from excretion of waste products that normally tolerant organisms such as crabs, have to cling to the pool margin, neither in nor out of the water. Being visible they become easy prey for the reef heron.

Immediately the tide returns conditions go back to normal marine conditions.

Pools toward the edge of the Long Reef platform support *Hormosira banksii*, southern sea lettuce *Ulva australis*, and common kelp *Ecklonia radiata.* Also common in larger pools are rockpool stars *Parvulastra exigua* (*Patiriella exigua*), the eight-arm seastar *Meridiastra calcar* (*Patiriella calcar*), snake-skin chiton *Sypharochiton pelliserpentis,* and several sea-urchins including the common sea-urchins *Heliocidaris erythrogramma* and *H. tuberculata*, along with the eastern egg urchin *Holopneustes purpurascens*.

Ulva australis with the shell grit anemone *Oulactis muscosa* with contracted tentacles.

Holopneustes purpurascens. The yellow suckers on the tips of the tube feet can be seen between the rows of spines..

Parvulastra exigua can often be seen during the day and is up to 15mm across.

Amongst the grazers are three turban shells, the Sydney turban snail *Turbo torquatus* characterized by two spiral ridges on the operculum, *T. undulatus* with a smooth green operculum and the largest individual the military turban snail also with a smooth operculum *T. militaris,* sometimes referred to as *T. imperialis. T. militaris* is at the southern end of its natural range at Long Reef. Changes in the populations of the different *Turbo* spp. may be indicators of environmental change driven by global warming. Another grazer common in pools is the elephant snail *Scutus antipodes* (*S. anatinus*) that has a very reduced flat white shell almost covered by a black fleshy mantle. Two other common grazers are the ribbed winkle *Austrocochlea constricta* (*A. obtusa*) and the zebra winkle *A. porcata* (*A. zebra, A. constricta*).

Sypharochiton pelliserpentis, note the snake skin-pattern on the girdle.

Three common intertidal gastropods, left to right: *Austrocochlea porcata, A. concamerata,* and *Nerita atramentosa.*

Pools are habitat to the gloomy octopus *Octopus tetricus,* numerous small shrimp and prawn-like animals that are quite difficult to identify and even the moray eel *Gymnothorax prasinus*.

Smooth and spiky forms of *Turbo militaris*.

Of the sessile (attached) species the compound ascidian *Didemnum moseleyi* may be seen along with anemones especially the shellgrit anemone *Oulactis muscosa* and the waratah anemone *Actinia tenebrosa*. Sponges are common,

Australian chiton *Ischnochiton australis* and an unknown shrimp, both are common under boulders in pools.

Gymnothorax prasinus normally hides during the day and moves out to feed at night. Photo by Dave Thomas.

particularly the purple encrusting sponge *Haliclona* spp., which have their surfaces covered in crater like holes. Another sponge of the same genus is greenish yellow and lacks the surface craters, and yet other sponges may be pink, liver brown, or orange-red.

And there is more to observe. The presence of fresh water from rain or seepage is necessary for some organisms but fatal to others. Certain animals rely on flowing salt water to bring their food and disperse their wastes. Examples are the many anemones of all sorts of colour, shape and size including the green anemone *Aulactinia veratra* (*Cnidopus veratra*), along with the waratah anemone *Actinia tenebrosa,* and the shell grit anemone *Oulactis muscosa.* These are generally highly visible but there are a few species adept at camouflage. Then there are other very much smaller animals, ones that look more like fine, wispy and fragile plants, with a large, 'flower' at the end of each branch. These hydroids are related to anemones, jellyfish, bluebottles and coral polyps. They include species of *Tubularia,* and the salt and pepper feather hydroid *Pennaria disticha* (*Halocordyle disticha*). Each 'flower' is an individual animal and it relies, as do most of its anemone cousins, on a constant flow of water that carries the food particles necessary for its survival.

Aulactinia veratra is normally found in flowing tidal gutters.

Actinia tenebrosa with tentacles open.

What all this indicates is that most organisms require an identifiable microhabitat that is suitable for their particular needs. A few hardy organisms are more tolerant and can exist almost anywhere but the greatest number find it essential to locate in their particular habitat in order to survive.

The rampart and bioconstruction features

Around most of the Long Reef platform edge and particularly well developed at the eastern end is a low rampart of saturated harder rock that stands slightly above the general level of the platform. Saturation, combined with bioconstruction (minor topographic features made by organisms) apparently limits rock weathering in this zone.

The species most important in bioconstruction are dense cunjevoi *Pyrura stolonifera*, encrusting Sydney coral worm *Galeolaria gemineoa* (*G. caespitosa* - the new name was applied when the original species was split using DNA evidence), and colonies of barnacles, such as a surf barnacle *Catomerus polymerus* and the rose-coloured barnacle *Tesseropora rosea*.

Fishermen and a patient pelican on the rampart at the edge of the shore platform, Long Reef NSW.

Galeolaria sp., densely packed in a bioconstruction feature on a wharf pile at Coffs Harbour, NSW.

Loose sediments

Boulder fields, scattered boulders, and small areas of sand occur at places around the headland, all providing habitat for different organisms. It is obvious that the sand in the beaches and particularly the thin spit of sand on the eastern end of the platform moves around with every tide and every storm. The boulders, however, appear to be immobile but appearances are misleading as comparison of air photographs taken in 1943 and satellite images of the present day clearly show that whilst the broad pattern of boulder distribution on the platform is much the same, no individual boulder can be seen in exactly the same place.

Boulder fields and beaches are fascinating places and the available living spaces in them depend on:

- the shape, stability, size, and spacing of the boulders and the cavities between them,
- whether or not they are embedded in finer sediment or layered on one another,
- whether they can be jostled by waves that may cause them to interlock or become imprisoned as they abrade one another,
- how many there are and how frequently they move or are renewed.

There are at least three subtle habitats associated with boulders: the upper surface, the under surface, and the substrate under the boulder. Large boulders may also have shaded and unshaded faces and can even have small cavities within them. Each surface has a different set of organisms associated with it and it is therefore very important that if you pick up or turn a rock over, please gently replace it as it was.

The tops of boulders or the upper surface often have a more luxuriant cover of foliose algae than the adjacent platform, as grazing pressure on these isolated and exposed surfaces is often lower. They are isolated little pockets or refuges from which rare species may be distributed after they have been depleted elsewhere. In this way they probably play important species-survival roles analogous to springs and waterholes in desert ranges where core populations of plants and animals may survive drought.

The under surface of boulders often support a range of sessile animals such as spirorbid worms, tube worms, keyhole limpets, bryozoans, sponges, hydroids, chitons, brittle-stars, crabs and the juveniles of more motile species such as a common sea-urchin *Heliocidaris erythrogramma* living in an isolated confined space where predators cannot easily reach them.

Side of a boulder showing a gradation from the algae surface through the pink encrusting coralline algae to the more sheltered habitat of chitons and echinoids.

The substrate is the material on which the boulder sits. It may be other boulders where the dark open cavities are crab paradise occupied by species tolerant of low light conditions and free water circulation. Rock crabs; *Ozius truncatus, O. deplanatus* and the red bait crab *Plagusia chabrus* are common here and you will often find their moulted carapace (shell) abandoned as they grow larger. However, the substrate is more often a sandy bed occupied by burrowing organisms. If the sand is densely packed and water circulation is limited, anaerobic conditions (no oxygen) that favour bacteria may develop and the body of the sand will be a drab grey colour. In muddy sediments sulphate reducing bacteria and other bacteria that metabolise methane will dominate the system. These conditions are rare on most shore platforms although the soft rock muddy shores of Auckland Harbour are an exception.

The animals found in the fine sediments are worms, crustaceans, and bivalves. Species diversity is low and ephemeral populations of species that can colonise the shifting substrate are present. Examples include isopods, amphipods, copepods, the ocean surf crab *Ovalipes australiensis*, the mauve sand shell *Austromactra rufescens*, the surf clam *Paphies elongata,* perhaps even the pipi *Donax deltoides*. These last two species are actually most common in the sand on open beaches.

Base of a boulder with a stony coral *Culicia* sp. Stony corals appear to be more common on Long Reef today than they were a few decades ago. Is this evidence of changes in water temperature and climate?

Base of a boulder with *Ischnochiton australis* and the mysterious chiton *Cryptoplax mystica*.

Crabs love the spaces between boulders, the red bait crab *Plagusia chabrus* at home.

Aspect

One of the important influences on the shape and size of any shore platform features and their biology is aspect, or the direction that any part of the rocky coast faces. On the east coast of Australia all aspects are represented, even a few westerly ones and although the subject is important we will only make a few points about it.

Aspect controls the amount of sun exposure and drying that a rock face may get, it affects the energy distributed by waves, and some aspects will be more influenced by warmer waters from the north or cooler waters from the south.

The big picture on the east coast is that sites facing north and east will have prevailing north-easterly winds for six months. On these aspects the cliffs may stand at flatter angles, a wedge of loose sediment often buries the base of the cliff, and boulders are more common. Waves and wind from this direction bring sub-tropical or tropical neustonic organisms (those living on the water surface) onto the coast, as well as the larvae of benthic organisms (those living on the bottom).

In the winter, the winds and rough seas come from the south together with a set of colder water organisms. South facing cliffs tend to be higher and steeper, rock falls are frequent but the debris is removed quickly and this results in a lot of abrasion on the platform surface as boulders are moved into deeper water. The shore platform is often narrower. With constant waves breaking on the shore the water is well oxygenated, therefore planktonic food resources are abundant and we might expect a high concentration of filter feeders, provided that they can hang on to the substrate during storms.

Westerly aspects, although rare tend to have drying winds off the land, a lot of sunshine and they tend to be in protected waters so wave energy is limited. These unusual conditions may favour rare organisms.

Moving from the big picture to the detail aspect is important in controlling the distribution of organisms on either side of a crevice or a boulder. Other things like rock type and structure being equal, aspect will dictate which side is drier, warmer, exposed to more sunshine, and different wind and wave conditions than another. It is fascinating to look closely at a rocky foreshore and work out not only how, but also where certain organisms live – often the answer is in aspect.

Bioerosion

This supposedly tired old world is really like a fickle human constantly remaking itself. Parts wear out, other parts are renewed. In human terms, this process seems unchanging, but in geological terms there is constant renewal. Rivers erode the land, volcanoes erupt new rock, glaciers scrape valleys clean, while winds move and deposit sand grains in ever deepening drifts. All the sediments formed by these processes are finally cemented together to make new rocks. This is all part of the rock cycle and geological erosion.

There is also a quiet process of biological erosion, or bioerosion that takes place, where nature destroys with one hand while building with the other.

Heliocidaris erythrogramma the species mainly responsible for creating 'urchin apartments'.

A sea-urchin appears rather formidable with all its spines, as it flaunts them in a showy means of protection, but sea-urchins are long-lived creatures, with the habit of homing in to one favourite rock hollow, day after day. Over time they make their hollow bigger. Sea-urchin spines and teeth (Aristotle's lantern) abrade the soft shale, and this incessant action results in the excavation of a hollow. Look in the rocky crevices and overhangs on the low tide edge of the shore platform and you will see 'apartment blocks' where generations of urchins have slowly etched away the bedrock. In time these faces collapse and the process starts again.

Some molluscs are also rock borers. Surprisingly, they are all bivalves, with apparently thin and fragile shells. They bore into rock or wood mechanically with a very slow rocking of their abrasive shells. The valves or shells of the delicate angel wing molluscs *Barnea australasiae*, have a series of sharp file-like spikes that do the damage. The animal moves its body (and shell valves) slowly excavating a hole bigger and deeper, until it is embedded and its two siphons are all that allow it communication with the outside world, for food intake and waste removal.

In their small way the gastropod molluscs add their bit to erosion. Some species of limpets will etch a shallow cavity into the surface of the rock using secretions and constant wear to enlarge their home-scar. Chitons have magnetite (iron manganese oxide) in their teeth (on the rasping tongue or radula), many limpets have combined iron oxide (goethite) with silica in their radula teeth and other gastropods use apatite, a mineral phosphate of calcium and fluorine. All of these are hard mineral structures that can actually scrape away rock while feeding. Grazing gastropods operate all over the shore platform and are probably a lot more effective in wearing away the rock surface than you would imagine.

'Aristotle's lantern' the mouth-parts of a sea urchin.

Sea urchin apartments worn into rock at the low tide edge of the platform at Long Reef, NSW.

Barnea australasiae in sandstone, Kurnell, NSW.

Grazing trails of *Nerita atramentosa*.

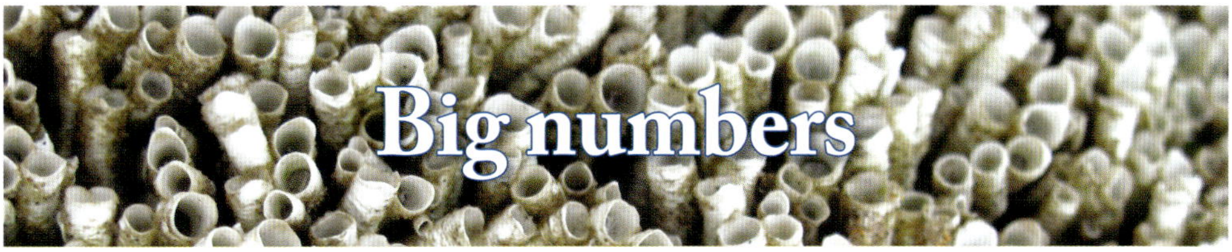

Big numbers

A reader of the draft of this book once said 'It's easy to see that a malacologist wrote it!' (a malacologist being a glorified shell collector.) This statement may be true, but perhaps it is also indicative of the fact that the Phylum Mollusca is not only one of the dominant phyla in the inter-tidal environment, but that it also shows how nature can exploit so many niches in so many ways.

World wide about 85,000 species of mollusc have been reliably described and some 8,700 are found in Australia. Of the total some 53,000 are marine and most of these are gastropods. This is but one class of molluscs with a single shell (univalves) that includes not only snails, limpets and cowries, but others that have lost most or all of the shell such as the nudibranchs and slugs. Soft parts of the body can usually be withdrawn into the shell, which is often sealed by an operculum. Most gastropods have two tentacles on the head with black eye spots. The mouth contains a ribbon of chiton (radula) with embedded file-like teeth (radula teeth) that are used for rasping food into small portions. The track of a feeding gastropod is often marked by distinctive wavy abrasion patterns on the surface, very visible in a glass aquarium or as the silver trail of a garden snail.

The gastropods are further classified into three subclasses, coastal representatives of these groups are:

Subclass Prosobranchia – the spiral shelled snail like creatures of the sea with a single pair of tentacles and usually separate sexes.

True limpets. *Cellana tramoserica* is a homing species with a distinctive pattern of light and dark bands often with a reddish tinge radiating from the peak of the shell. *Scutellastra chapmani* has a distinctive star shaped shell with eight rays. It lives at the lowest levels of the tide and is common as a dead shell in loose sediment.

Distinctive eight point star of *Scutellastra chapmani* one of the true limpets consumed by oystercatchers and turnstones.

Scutallastra peronii a species named after Francois Péron the naturalist on the Baudin expedition from France that spent five months in Port Jackson in 1802.

Keyhole limpets. *Diodora lineata* has a distinctive hole at the apex of the shell. When crawling the animal appears to be much larger than the shell. Found in the low tide zone, often under stones.

Scutus antipodes and a chiton in a rolled up defensive position. The limpet attached to the *Scutus* is probably *Patelloida mufria*.

Slit limpets. The elephant snail *Scutus antipodes* has an indistinct notch in the front margin of the shell. The notch is more obvious in other species. Found at mid to low tide amongst the cunjevoi and *Galeolaria*, the distinctive black mantle almost completely covers the shell.

Ear shells. Better known as abalone *Haliotis* spp., are limited to sub-tidal water and have been heavily predated by humans. Very occasionally found in low tide joint crevices.

Amongst the spiral shelled gastropods the following examples are common:

Trochids or top shells such as the strongly ribbed *Austrocochlea constricta* found at higher tide levels.

Turban shells are large snails with solid calcareous opercula. *Turbo militaris* is the largest but has tropical affinities; *T. torquatus* is up to eight centimetres across with a heavily sculptured operculum and *T. undulatus* is the smallest and has southern affinities.

Nerites The black nerite *Nerita atramentosa* is very common around the southern coast and is found in dense clusters from mid to high-tide levels.

Littorinids are well represented by *Nodilittorina pyramidalis* with its distinctive whorls of surface nodules found at the highest tide level. Slightly lower is the pale blue *Austrolittorina unifasciata* (*Littorina unifasciata* and *Nodilittorina unifasciata*) and at low tide these are replaced by zebra winkles *Bembicium nanum* amongst the green algae.

Nodilittorina pyramidalis one of the common littorinids.

Austrolittorina unifasciata another common littorinid.

Bembicium nanum.

Sand snails or moon snails such as *Conuber sordidum (Polinices conicus)* which is carnivorous and lays a 'sausage jelly' egg mass that is frequently seen on the shores of muddy inlets.

Cowries are a largely tropical family with about 30 species in NSW. Not often seen but easily recognized, they are usually found among the rocks at low tide. The distinctive feature of cowries is that the animal's mantle covers the outside of the shell keeping it clean and shiny. Cowries feed on algae, sponges, dead organic matter and other molluscs.

Variation in pattern on *Monetaria caputserpentis* with a juvenile shell at the top left, Long Reef, NSW.

Tritons or trumpet shells are all predatory carnivores represented by Spengler's triton *Cabestana spengleri* which preys on cunjevoi and the Australian red triton *Charonia lampas* which feeds on echinoderms. Remains of these species often turn up in Aboriginal middens.

Charonia lampas, note the horny operculum.

Murex shells The cartrut shell *Dicathais orbita* is related to the species that produced Rome's Tyrian purple dye and is found in groups, often adjacent to their egg capsules in wave swept crevices at low tide level. In different environments this shell has considerable variation in the degree of ornamentation (ruts). The species can live to 19 years but the average is about five. The mulberry snail *Morula marginalba* also in the family Muricidae, is found right across the shore platform.

Dicathais orbita with prominent ruts on the shell, some examples can be quite smooth.

Mitre shells Elongate mitre shells are mainly tropical species but a small brown example *Isara carbonaria* may be found at low tide level in crevices or under rocks. It is fond of peanut worms.

Whelks The most common local example is probably *Nassarius glans particeps* a scavenger with a distinctive black-spotted foot.

Nassarius glans particeps possibly a southern variant or sub-species of *N. glans*.

Cone shells Cone shells are also a tropical group with *Conus papilliferus.* and *C. anemone* being two southern representatives found under stones at low water level. *C. papilliferus* has a variable colour pattern in that its brown patches may totally cover the shell or be almost absent. All the cone shells are carnivorous, feeding on worms, molluscs or small fish. They have a highly specialized feeding mechanism where the radula teeth have been modified to hollow harpoons. When prey is sensed, a harpoon is filled with poison and attached to the proboscis. The prey is speared and drawn to the mouth. The poison in some tropical species has caused human fatalities and although temperate species are usually regarded as harmless they should be should be treated with caution.

Subclass Opisthobranchia – gastropods with two pairs of tentacles, much reduced shells or even no shell at all. They are often hermaphrodites.

Bubble shells like *Hydatina physis* and *Bullina lineata* have very thin shells and although usually found in estuaries and sheltered bays they may be washed up along the shore during storms and are occasionally common on shore platforms. *Hydatina* has a very attractive fluted brown mantle with an iridescent blue edge. Another quite rare species in NSW *Micromelo guamensis* looks very like the more common *B. lineata* but is actually more closely related to *H. physis*. There is a question as to whether or not this is the same species as *M. undata*.

Hydatina physis with eggs.

A pair of *Bullina lineata*.

Micromelo guamensis normally distributed in Japan, South Africa and Western Australia was found found on Long Reef, December 2010. Photo by Peter Miller.

Sea hares belong in deeper water but migrate to the rocky shore for breeding. Some species exude a purple dye when disturbed and lay long-tangled strings of orange eggs amongst the seaweed. Although appearing to have no shell *Aplysia argus* and *A. sydneyensis* actually have a thin horny shell just visible under the mantle. *Dolabella auricularia* and *Dolabrifera brazieri* both have an internal shell of lime. Sea hares are often seen grazing on the walls of coastal swimming pool.

Dolabrifera brazieri a common sea hare on the Sydney coast but hard to photograph.

Nudibranchs are snails with no shellthat are often mistaken for others but all have a characteristic cluster of branched feather-like gill plumes that emerge from the back. In shallow water they include; bright pink short-tail nudibranch *Ceratosoma brevicaudatum*, the orange spotted *C. amoenum* that is also found in New Zealand, the sky blue *Hypselodoris bennetti* with a bright yellow mantle, and the white to lemon yellow *Doriprismatica atromarginata* with a fine black mantle edge.

Ceratosoma amoenum. Photo by Dave Thomas.

Doriprismatica atromarginata. Photo by Dave Thomas.

Sea lizards or aeolids are the bizarre looking *Glaucus atlanticus* and *Glaucilla marginata*. Both creep along the under-surface of the open ocean upside down whilst feeding on coelenterates such as *Velella velella* and *Physalia* spp. These are occasionally stranded in large numbers on our shores.

Glaucus atlanticus the bright blue upper surface is actually the ventral surface. The dorsal surface is underneath and a lot paler, an example of counter-shading,

Subclass Pulmonata gastropods in which the mantle space acts as a simple lung. Not surprisingly these organisms are capable of surviving long periods of exposure to the air.

False limpets represented by *Siphonaria funiculata* a cone-shaped limpet like form found in large numbers at higher tidal levels and *S. diemenensis* that has a more flattened shell with conspicuous radiating ridges. Both are found amongst the true limpets right through the tidal zones. *S. diemenensis* is the only limpet like animal that you can flick off the rock with a quick fingernail.

Siphonaria diemenensis, easy to detach but if the behaviour of oyster catchers can be relied on they may not be so tasty.

Before too much credit is given to the Phylum Mollusca, we should also look at the Phylum Crustacea, which in the marine environment encompasses a huge group of arthropods (animals with segmented bodies and jointed legs).

The molluscs have giant clams and huge squid in their flagship. The crustacea have lobsters and large crabs. But in both phyla it is the smaller species that play dominant roles.

Turn over a rock at the low tide level and you may find 50 different species of molluscs. An equivalent volume of water might hold as many different crustaceans in the plankton, free swimming organisms going under the fancy names of copepods, amphipods, isopods etc., most of which grow to adulthood in that plankton. The same plankton will also harbour juveniles of crabs and lobsters. The stranded and decaying seaweed on the beach might contain myriads of minute hopping and writhing crustaceans, doing what they do best, ingesting and breaking down dead kelp to help recycle nutrients. So in size variation, number of species, diversity and capacity to exploit every available niche, both Crustacea (about 7,300 Australian species) and Mollusca (about 8,700 Australian species) are excellent examples of the capacity of nature not only to exploit habitats, but also to create them.

Early 20th Century illustration of variety in the decapods. Ernst Haeckel, 1904 Kunstformen der Natur. Plate 86. Source:Wikimedia Commons

Synergies in life

In the terrestrial world we are well aware of certain organisms that have a remarkable relationship or affinity to another, often quite different, organism. Some relationships are friendly, like epiphytic orchids on trees. Some are not, like parasitic heartworms in dogs, or lice on humans.

In the marine world these relationships have been finely honed and there are as many combinations and permutations as could be imagined although it can sometimes be difficult to work out which species gains, which loses and exactly how they interact. The relationships can be broadly divided into four main categories. There are myriad examples of these relationships, and one could write a book about them (surprise, surprise, it has been done) but we will only select a few examples that can be seen on shore platforms if the eye is keen.

An example of aegism, *Cellana tramoserica* with free-loading barnacles that are out of reach of *Morula marginalba*.

Aegism (*aegis* = protection; the shield of Zeus) is, as the name implies, a life style that offers protection to one partner, without necessarily providing a mutual benefit to the other. It is very common, ranging from one organism living on another to exploit a surface; barnacles on limpets, bryozoans on kelp; to a degree of specific reliance of one species on another for protection with no hint of food sharing; a fish 'borrowing' the temporary protection of a snapping shrimp hole to avoid a predator, or certain fish species utilising the protection offered by sea-urchin spines; through to a permanent relationship between two organisms, such as certain barnacles living on and specific to whales.

Commensalism (eating at the same table) expresses a shared gastronomic interest in that the smaller partner usually consumes the scraps left by the larger organism. There may be other benefits to one or both partners, but they are incidental. Most associations of this nature have one partner showing cryptic habit, so that visible examples are few and far between. Certain shrimps are always to be found within bivalve shells; bivalves of the Superfamily Galeommatoidea often seem to be found commensally with some other organism and have probably exploited this phenomenon more than any other group. These generally small to tiny bivalves are to be found in worm tubes, attached to certain crustacea, on other bivalves, in mantis shrimp holes, associated with retractile anemones, and are even found internally in some holothurians, bivalves and sea urchins.

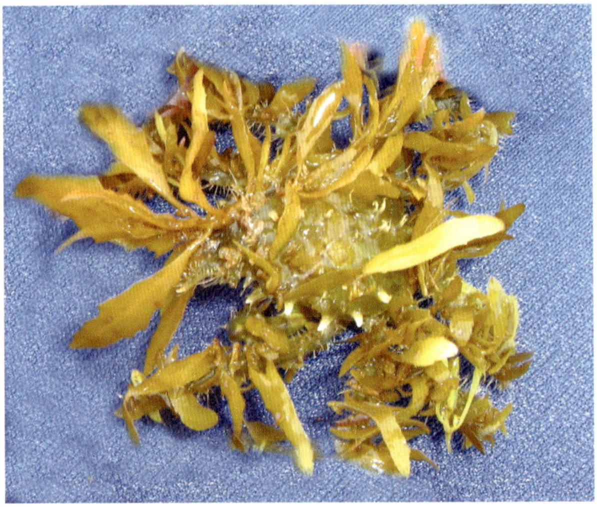

An example of mutualism, a decorator crab probably *Naxia tumida* with excellent seaweed camouflage. The strands of seaweed are hooked onto tiny curved spines on the carapace that works just like Velcro.

Mutualism (both sides benefiting equally) often occurs between species from unrelated phyla and shows intimate relations between species usually at war with each other. We find cleaner shrimps removing parasites and cleaning lesions within the mouths of fish or eels, anemone fish living safely within the stinging tentacles of anemones, with the fish bringing food and the anemone providing protection. Common visible examples are generally not available on our shores, but one good example is that of hermit crabs and anemones that live attached to the crab's borrowed shell. As the crab moves to larger premises, it carefully 'transplants' its anemone companions on to the new shell. The benefits to either are not immediately clear but presumably the crab gains camouflage and the anemone gets to move around and probably accesses food scraps missed by the crab.

Parasitism (winners and losers) is a common occurrence and this is when the parasite has a negative effect on the host. The parasite is often small and cryptic, beautifully camouflaged, lives internally, or is otherwise not obvious. Local marine examples are minute gastropods of the family Eulimidae parasitic on sea-stars, and blood sucking isopods on the skins of fish. The most successful parasite is the one that relies totally on its host for nourishment but does not kill it.

Bigger ideas

All of the above only explains some of the more particular relationships between one or two species at a time. But on the shore platform we are really looking at a myriad of organisms and an incredibly complex ecosystem about which ecologists continue to develop models to improve understanding. Inevitably our models are simplifications of reality but some are more durable than others and do provide us with tools that can improve resource management.

One long-standing approach has been to construct food webs for a few species of interest and to link all the known feeding relationships into a diagram or a semi-quantified computer model. All textbooks have examples of these food webs but although they serve to make us aware of trophic pattern as we make more and more observations they can become so complex as to be unworkable.

And another approach, or is it?

In recent years a new approach was adopted with some enthusiasm. This was the idea that certain 'keystone species' were present in most communities and ecosystems. It began with the observation that some predators and grazers seem to have such a strong influence on the presence or absence of other species in a community that they were thought to be responsible for local biodiversity. Animals with this influence appear to play a dominant role on the overall 'stability' of an environment out of proportion to their own small numbers and biomass. They were referred to as keystone species because if they were removed then the composition and structure of the community could pass through a cascade of change. There may be 500 organisms in a given area. Remove any one of 495 and nothing seems to happen. Remove any of the other five and a significant change can follow. The concept is supported by experimental data and it looked as if it should have management value, but despite much research it has never been easy to identify keystone species without actually removing the suspects and many ecologists now think that the concept is too fuzzy to be of practical use.

An example of a keystone species can be seen in the recent demise of kelp forests around the world. The kelp species involved were critical in shaping the physical structure of those marine forests but were sensitive to water temperature. When temperature increased, the forests collapsed and the structure and composition of the entire ecosystem shifted. A second example may be our mulberry snail which is such an effective carnivore that it appears to keep numbers of many other shellfish in check.

Pattern between the tides

Zonation on a vertical rock face from *Galaeolaria* sp. to barnacles, Nundera Point, NSW. The vertical interval in this case is less than 1 m.

Within the ocean the environment is relatively constant and benign except for storms. On the NSW coast water temperature ranges between 15 and 20°C from June to November and between 21 and 27°C from December to May. Salinity of the open ocean is 35 parts per thousand (3.5%) but near the shore this can drop for short periods when heavy rain occurs and stormwater drains are in flood. Open water pH is close to 8.0 (mildly alkaline), light penetration is good, oxygen, nutrients, and food supply, are all readily available. The water provides an excellent growth medium for attached algae and filter feeders, and is a supportive environment for creatures that float, swim or crawl.

The tidal zone is a variable intergrade and often a stressful environment but there are some general patterns. For example, on the rocky shore as you move down from the spray zone (above high tide) to total immersion (below low tide):

- species diversity increases, with the greatest variety being found at the low tide level;
- the area of the surface occupied by organisms increases;
- habitat structure and complexity increase;
- biomass increases; and
- total productivity increases.

These variations are seen in a common pattern called zonation where different ‘indicator’ organisms are most abundant at different levels of wetting (tide range plus wave action).

Zones or Patches?

Whilst it is common to recognise 'zones' on shore platforms the use of this term is contentious and not everyone accepts its traditional meaning. The alternative view of the distribution of organisms is one that emphasises the more random pattern of 'patchiness' where the entire inter-tidal community can be seen as a mosaic of assemblages or succession stages from bare rock through to a complete cover by dominant species. Patches change quickly and small patches tend to be more dynamic than large ones.

Zonation on a wharf pile cunjevoi and *Galaeolaria sp.*, Coffs Harbour, NSW.

Like many academic arguments the merits of zones versus patches are not easy to resolve because it involves two approaches to the same phenomena. Zonalists tend to be observers of natural history, whereas biologists or ecologists advocating patchiness favour approaches that are strongly experimental and that can be modelled. In reality both approaches have value, with zonation giving us an impression of pattern stability and longevity, and patchiness emphasising the processes of change.

A patch in cunjevoi and *Galaeolaria sp.*, apparently kept open by limpets, Towradgi Point, NSW.

Zonation is best seen on steep shores with moderate wave energy. In storm affected areas selection processes favour species better able to hang on, and in quiet waters the reverse is true. But in moderate wave energy zones representatives of all groups are likely to be present. Patches are probably best seen on wide flat, low-energy platforms where the zones that are so prominent on a vertical cliff face or a wharf pile, are smeared out and more extensive.

The common organisms found in the different zones change as you travel along the coast into the tropics, or south to colder waters. Cunjevoi *Pyrua praeputialis* for example are not found beyond Southern Queensland or west of Cape Otway. The West Australian species is now identified is now as *Pyura dalbyi*. At Bermagui bull kelp *Durvillaea potatorum* suddenly appears and becomes more abundant than common kelp and it can be found around Victoria and Tasmania. This bigger pattern can be linked to the different coastal water bodies and their temperature but wherever you are there are always zones.

The terms that are used to describe zonation vary with different authors who sometimes subdivide the zones, but we have opted for the simplest split as we see it along our temperate coasts.

Inter-tidal zonation

Terrestrial zone - land above the tide limit dominated by land plants but exposed to salt blown in on the wind.

Supralittoral – the splash and spray zone above high water springs where algae, lichens and a few periwinkles are found.

Littoral – the inter-tidal area between high water and low water springs. It is usually split into three parts:

Upper littoral – dominated by barnacles

Middle littoral – dominated by Sydney coral worm

Lower littoral – dominated by cunjevoi

Sublittoral – where marine conditions occur and kelp is dominant.

The terrestrial zone

The terrestrial zone is best defined as the edge of the land where there are no living marine organisms. The adjacent sea can affect it as it may be a rocky cliff, or covered in dune sand, and it receives a large amount of salt from breaking waves. Only hardy plant communities survive because salt spray burns the leaves and plants are wind-pruned into wonderful geometric shapes that look almost like those in a formal Japanese garden.

The common plants found here differ from south to north along the NSW coast but around Sydney the coastal woodland or headland heath includes coast banksia *Banksia integrifolia,* coast tea-tree *Leptospermum laevigatum,* coast rosemary *Westringia fruticosa,* and coast wattle *Acacia longifolia* subsp.

sophorae. Areas of low heath and kangaroo grass *Themeda australis* also occur. Drooping she-oak *Allocasuarina verticillata* (*Casuarina stricta*) are limited to headlands on Narrabeen Shales and swamp oak *Casuarina glauca* occurs on the water's edge in the estuaries and coastal lagoons. On sand at the edge of the beach and at the base of the cliffs you are likely to find hairy spinifex *Spinifex sericeus* (*S. hirsutus*), coastal pig-face *Carpobrotus glaucescens*, beach fescue *Festuca littoralis* (*Austrofestuca littoralis*), spiny-headed mat-rush *Lomandra longifolia*, sea rocket *Cakile edentula* (*C. maritima*), and coastal jack bean *Canavalia rosea* (*C. maritima*).

Flower spike of *Banksia integrifolia*.

The terrestrial zone dominated by lichens and wind shear across all higher plant species on the coastal fringe at Nundera Point, NSW. A zone of high salt accession.

To the north rainforest species occur in the dunes or as stunted plants on the headlands. Common species include tuckeroo *Cupaniopsis anacardioides* and red ash *Alphitonia excelsa*. To the south cool temperate species such as blackbutt *Eucalyptus pilularis*, red bloodwood *Corymbia gummifera* (*Eucalyptus gummifera*) and spotted gum *Corymbia maculata* (*Eucalyptus maculata*) are more typical.

Unfortunately there are also many introduced weeds. Perhaps the most widespread and troublesome are bitou bush *Crysanthemoides monilifera* along the southern half of the coast, and lantana *Lantana camara* in the north. Bitou bush was introduced from South Africa to control drifting sand. It was widely used by government agencies and sand miners even as late as the 1970s and now we spend time and effort trying to control it. Marram grass *Ammophila arenaria* was introduced from Europe to control sand drift along the southern NSW, Victorian and Tasmanian coast and although we don't usually think of this as a weed it changes the shape of the dunes and must have other ecological impacts but these have not been studied.

Spinifex sericeus. This plant has no link to the porcupine grass or 'spinifex' *Triodia irritans* and others of the desert sand dunes.

When is a plant a weed? When we deem it to be so!

The supralittoral

The supralittoral is above the high water mark where small gastropods generally known as periwinkles, in the family Littorinidae live. This zone is the interface between the sea and the land. It receives a lot of salt spray and is typically nearly bare rock. Weather conditions dictate large variations in salinity, temperature, and wetting. There is limited access too, with king tides being necessary to deposit free-swimming larvae, which then have to evade the whole gamut of predators. As the snails grow they climb higher.

Around the world the supralittoral zone is an 'arid' environment that has a low diversity of species and is typified by a 'black line' of lichens such as *Verrucaria* spp., just below a zone of orange lichens such as *Xanthoria* and *Caloplaca* spp.

Very few marine species are found here although three periwinkles are common; the blue periwinkle *Austrolittorina unifasciata,* and the nodular periwinkle *Nodilittorina pyramidalis* a warmer water species, and the zebra winkle *Bembicium nanum,* a species that varies in shape from tall and conical in quiet water to almost flat in stormy areas. Two other gastropods in this area are the black nerite *Nerita atramentosa,* and Petterd's limpet *Notoacmea petterdi.*

Austrolittorina unifasciata is usually found in dense clusters while exposed during low tide and *Nodilittorina pyramidalis*, a periwinkle with a knobbly shell that has the same diet can be found as much as 10 metres above the high tide limit, also clustered in pits, crevices or hollows. This snail is remarkably tolerant of high and low air temperatures. Why are these species generally separated in level and why do they cluster? Why does one have a knobbly shell and one a smooth shell?

One suggestion to explain clustering is that grazing snails tend to move higher on the shore because competition is less and food is more abundant. But that explanation needs to consider what the competition is, and it needs to be offset against the reduction in available grazing time, the increased exposure to desiccation, and predation on the drier and higher shore levels.

Three species of lichen on coarse-grained essexite with the large yellow/green (probably *Xanthoria* sp.), showing typical concentric growth pattern, Nunderra Point, NSW.

In the case of *A. unifasciata* details of the distribution make the story even more intriguing. These snails are not found below the mid-tide level. In the highest density clusters (nearly 10,000 individuals per square metre) at about high-water-springs, all of the individuals are small, and the largest snails are found either very low in the range or very high in the range. The species does not appear to be eaten by anything much. Mulberry snails *Morula marginalba* take some at the low end of the range but although crabs and birds may be predators higher on the shore there are few observations to support this. What does change through the range is food supply. At the low end of the range diatoms and juvenile algae are reasonably abundant and as *A. unifasciata* is able to graze for longer periods of time we can speculate that the snails grow quickly. At the top end of the range the rock surface is covered by lichens, but grazing can only be done when it is wet. Nevertheless there are so few snails here that food is not limiting and individuals can grow quite large although they may be much older. In the middle of the range where the dense clusters occur it seems that crowding is so great that food is limiting and these animals are on the verge of starvation, therefore the average size is reduced and mortality increased.

Hatchling and juvenile snails are much more sensitive to desiccation than adults and survival is better lower on the shore platform. They are also more abundant than adults and presumably are subject to much higher levels of predation. Life is tough when you are little and evolution has sifted strategies to cope with these conditions. Success comes to those which manage to settle in favourable protected microhabitats such as patches of filamentous algae, amongst mussel clusters, or within dense aggregates of barnacles. These locations not only limit desiccation but also provide more food. Getting there is the trick, as juveniles are initially subject to wave dispersion and when settled they move very slowly. So where is the best place to live if you are a blue periwinkle? Perhaps it depends on chance. Or do even blue periwinkles make their own choices?

As for the knobs on *Nodilittorina pyramidalis* we suggest that because knobs increase the surface area of a shell they may function to help cool the organism when it is exposed to high air temperatures. On the other hand older texts would have simply referred to the knobs as 'decoration' and perhaps not all variations in form and pattern actually need to have a practical function?

The littoral

The littoral (sometimes called the eulittoral) is the inter-tidal area between high water and low water springs. This is the zone of greatest biological diversity. Barnacles of several types along with grazing snails mark the zone and three sub-zones can be identified;

The upper-littoral is always dominated by barnacles especially the honeycomb barnacle with four side plates *Chamaesipho tasmanica*, and the eastern shore barnacle with six plates *Chthamalus antennatus*. Incidentally *C. antennatus* is one of many barnacles originally described and named by Charles Darwin in 1854. He spent eight years working on barnacles all over the world and published four volumes on both the living and fossil representatives and a great many of his names still stand.

The middle-littoral is represented by the rose-coloured barnacle *Tesseropora rosea* which reveals lines of pink shell within strongly fluted (eroded) wall plates, and a surf barnacle *Catomerus polymerus* with its extra whorls of basal plates. Both prefer vertical rock faces and can spread high up into the surge zone. The lower part of the middle-littoral can be occupied by thick crusts of the Sydney coral worm *Galeolaria gemineoa* that may appear as a solid white band around the rocks.

The lower-littoral is often crowded with massed cunjevoi *Pyura praeputialis* growing from a mat of tufted coralline algae *Amphiroa anceps* and *Corallina officinalis*, and red seaweeds, amongst which the giant chiton *Plaxiphora albida*, and the eyed chiton *Onithochiton quercinus* can be found grazing.

Tesseropora rosea, this barnacle may be found as far away as the Kermadec Island (northeast of NZ), New Caledonia and South Africa.

Within the whole of the littoral zone limpets are common and these are one of the most important controls on the abundance of algae. The patellid limpets such as a slit-limpet *Montfortula rugosa* have such strong radula teeth that they can actually abrade the rock surface whilst feeding. However, false or siphon limpets such as *Siphonaria diemenensis*, and the denticulate siphon shell *S. funiculata* var. *virgulate,* have the capacity to breathe air, and feed on softer, larger seaweeds by biting off pieces.

Cunjevoi were once a favoured food of Aboriginal people and are still collected as bait by rock fishermen. Cunjevoi are sessile (attached) organisms found at the lowest tide level and into the shallow marine zone. They are known to visitors as sea squirts because if touched or disturbed when exposed at low tide they squirt a quantity of water into the air from one of two body siphons on the top of the leathery coat (tunic and hence tunicate).

Adult cunjevoi stand about 150 to 200mm high and in rough water conditions can be very crowded on the low tide rocks. They are not very attractive to look at, are often coated in seaweed, and a lot of people imagine them to be plants. In fact they belong in the Phylum Chordata because their free-swimming larvae have a notochord that is a precursor of a vertebrate spinal chord. The larvae have a short life as free individuals, maybe 24 to 36 hours. They apparently detect the presence of adults from chemical signals in the water and attach themselves head down in an existing colony. There they spend the rest of their life, probably a few years, functioning like a stationary vacuum cleaner pumping seawater through their bodies to filter feed, to obtain oxygen, and to dispose of wastes.

Amphiroa anceps alive, note the jointed 'elbows' that provide flexibility.

Bleached *A. anceps* as commonly found on the beach.

Pyura praeputialis extends from Shark Bay on the Western Australia coast, around the south coast and Tasmania, and up past Sydney to Noosa. If adult cunjevoi cannot move and the larvae have such a short period of free movement then how does the population maintain a single gene pool across such a great distance and in different water bodies? Another puzzle for future research.

Dense colony of *Pyrura praeputialis*, an open coastline that is subject to high wave energy, Towradgi Point, NSW.

Cunjevoi had long puzzled people as to whether they are animals or plants until 1866 when a Russian scientist completed a life cycle study. This showed that the 'tadpole' of the free-swimming larval stage has a rudimentary backbone (notochord), brain, and an eye spot. These characteristics place them in the Phylum Chordata making them more closely related to humans than any other invertebrate, though you would never see that when looking at the adult.

There are five species of *Pyura* in Australian waters and on a global scale our cunjevoi, formerly identified as *Pyura stolonifera* has a strange disjunct distribution. It is found on the Atlantic coast of North Africa, around South Africa, in southern Australia, on the northern tip of New Zealand and near Antofagasta in Chile. DNA analyses have shown the New Zealand and Chilean populations to be genetically identical to cunjevoi in eastern Australia which we now recognise as *Pyura praeputialis*. Both countries have had trading connections with Australia and the transfer of adults as hull fouling, or perhaps as larvae in ballast water are likely explanations. Does it matter? Perhaps yes, as cunjevoi form dense colonies and they may exclude endemic organisms. It is certainly too late to control them and time will tell.

The sublittoral zone

This is the zone where marine conditions occur. It is marked by the upper limit of kelp that is only rarely exposed at extreme low water springs in calm conditions. It is a fairly stable environment except for variations in wave conditions and weather. However, there is potentially greater predator access at all times, necessitating escape, camouflage or other avoidance strategies, but coupled with that, larval dispersal ability is always available. The sublittoral zone is dense

Phyllospora comosa is less common than it might be along the NSW coast because of heavy grazing by echinoderms.

The east Australian form of *Pyura gibbosa* extends to 30cm in length.

with brown algae and an undercover of encrusting coralline red algae. Common species include, broad-leafed sargassum *Sargassum fallax* near the surface, crayweed *Phyllospora comosa* appearing at the low-water mark, and common kelp *Ecklonia radiata.* As to animals a surf barnacle *Austrobalanus imperator*, and ascidians such as the sea tulip *Pyura gibbosa* are present along with a common sea-urchin *Heliocidaris erythrogramma* and the eastern slate-pencil urchin *Phyllacanthus parvispinus*.

Phyllacanthus parvispinus wedges itself into crevices during the day and emerges to feed at night.

Several species of mussels occur in the sublittoral zone and exploit hard substrates because they can attach so well with their shock-absorbing byssal threads. Mussels are one group of organisms about which we do have a reasonable knowledge of their biology because many are important commercial species. Bivalve file shells such as *Lima nimbifer* and *Limatula strangei* may be found with a diligent search.

Lima nimbifer a swimming bivalve that is difficult to find.

Limatula strangei a bivalve found under rocks and that can swim by pulsing its valves.

The marine zone

The zonation continues with depth to where the environment becomes totally marine, and fish are a full time component of the community. Changes in dominant algae occur with depth in response to the decreasing penetration of sunlight into the water and submarine conditions take over where the light cuts out at about 100m, but this is well beyond our level of observation from the rocky shore and the scope of this book.

The water surface

Yet another zone is the sea surface itself where a group of animals known as the neuston live. The surface of the water is a difficult place in which to survive. With exposure to scorching sun or driving rain, at the whim of wind and currents, those few organisms that occupy it are extremely well adapted to do so. Most species are counter-shaded blue. Viewed by predators living in the water below, these animals are a silvery blue, the colour of the water surface. To winged predators in the air above they show the deep blue of the open ocean.

Living on the water surface requires an animal to have some means of staying there and the easiest way is to produce a bubble raft or to ingest air. We don't usually see these species on the water but often find representatives of them stranded on the beaches and shore platforms. Two bluebottles are found along the NSW coast in summer. The limits of their distribution are not clear but the most common one around Sydney seems to be *Physalia physalis.* A related species *P. utriculus* occurs in cooler water further south. There are also two other less common related animals, the deep blue by-the-wind sailor *Velella velella*, oval organisms with a sail placed at a tangent, and the Porpita sailor *Porpita porpita*, with a deep blue disk about 2 centimetres across with lots of tentacles. All of these belong to the Phylum Cnidaria. They feed on animals in the plankton, with the bluebottles being capable of catching small fish, and the other two harvesting zooplankton.

Physalia physalis stranded on the beach with the tentacles contracted to a dark blue pulpy mess.

The bluebottle is not a single animal but a whole complex of individual zooids. The individuals cannot exist alone, only as the super-organism. One zooid is a loner, the pneumatophore that produces the bubble, and generates the gas that fills it. Three other large groups of zooids are suspended from it: the dactylozooids that exist in long strings with many thousands of stinging cells; the gastrozooids that feed on the animals caught by the dactylozooids and distribute the nutrients to the super-organism; and the gonodendia that are the reproductive zooids. The whole animal is a hermaphrodite, but produces males, which remain attached, and minute females, which float off and eventually produce another bluebottle by a process of budding.

Although largely at the mercy of wind and currents, the bluebottle does have a bit of manoeuvrability. It has been observed rolling over, or at least it has the capacity to immerse or expose different parts of the bubble, probably to avoid drying out and becoming brittle. Different colonies have their pneumatophores oriented to the left or the right of the main body and this means that they drift in different directions under the same wind. With an onshore wind half of any bluebottle raft may be driven ashore but the other half will sail out to sea.

Like many other cnidarians, there is a close commensal relationship between the bluebottle and other organisms, in this case several species of drift fish in the family Nomeidae. It is mainly the young fish that have this relationship, where the fish get protection at no expense or benefit to the bluebottle, and as they have immunity to the stinging cells they can forage whilst mingling. The association is so close that drift fish have been observed swimming aimlessly in the wave swash zone seeking their stranded invertebrate brethren.

Feeding on the bluebottles are several molluscs. Three are the violet snails, *Janthina janthina, J. umbilicata,* and *J. prolonga*, and a related rare pale brown snail *Recluzia hargravesi.* All of these species are supported by a bubble raft secreted from the foot and are washed up on the beach in summer. The other bluebottle predators are two marine slugs, though that seems a poor name for two beautiful and similar animals called glaucids or sea lizards, *Glaucus atlanticus* and *Glaucilla marginata.* Come to think of it 'sea lizard' is also a pretty poor name for creatures that are actually molluscs that lack a shell. Both are active bluebottle predators and have the extraordinary ability to ingest stinging cells from the bluebottle and use them for their own defence. Exactly how they do this and whether it is the stinging cells or toxins associated with the cells that protect them is still being debated a century after the fact was first determined. Be wary handling these if you come across them.

Another sea slug that has the ability to steal functional components from other organisms is the blue dragon *Pteraeolidia ianthina. S*ometimes found near Sydney it grows to between 12 and 18cm and has evolved to harness solar energy. Juvenile dragons are white, but adult dragons are brown with lovely blue to blue green cerata. This difference in body colour is because adults have somehow transferred microscopic plants (zooxanthellae) from hydroids into their own body. The zooxanthellae photosynthesize and some of the energy is available to the dragon. Solar powered dragons? Well that's one way to earn a living!

The blue dragon *Pteraeolidia ianthina.* Photo by Dave Thomas.

Plankton

The word plankton comes from a Greek word *planktos*, which refers to a wandering or drifting mode of life, a wonderfully apt word for an extremely important part of the marine system. Larger organisms such as fish that can move independently are called nekton, which is another Greek word meaning 'swimming'.

You might well ask, 'What have plankton to do with the shore platform?' That's a good question, as most of it is difficult to see without a good microscope. However, plankton contains the gametes and juveniles of a great many of the shore platform species. For these species the planktonic stage of their life cycle is the only time that they can actually move. This provides the opportunity for dispersal of genetic material from headland to headland and even across ocean basins.

Plankton are divided on size and whether or not they are photosynthetic. The smallest plankton are mainly bacteria. The next largest, but still less than 20 microns in diameter, include the phytoplankton, that are capable of photosynthesis. Examples are the single celled diatoms and radiolarians

Scanning electron microscope image of a foram *Elphidium macellum*, from the sand of Long Reef beach. This species has a global distribution. Photo by Tom Brougham.

that have skeletons (tests) made of silica, and most of the dinoflagellates. Phytoplankton convert sunlight into energy through photosynthesis and produce oxygen. On a global scale the total oxygen output from phytoplankton is about 65% of the oxygen that we breathe.

The largest organisms are mainly zooplankton. This includes many of the forams the empty tests (shells) of which can be found in shelly beach sand as well as larval components of many shore platform species. Comparison of planktonic forams collected in the tropical Pacific Ocean during the HMS *Challenger* expedition of 1872/76 with modern collections from the same locations has revealed a worrying decline in test thickness (as much as 70%) that is believed to be caused by ocean acidification. Quite large organisms are also part of the plankton such as bluebottles *Physalia* spp., jelly blubbers *Catostylus mosaicus*, and the moon jellyfish *Aurelia aurita.*

Larval components of zooplankton are adding all the time. They often bear no resemblance to the adult organism and linking the larvae with the adults and unravelling the life history of shore platform organisms has presented major challenges to biologists.

Along the NSW coast plankton numbers are lowest in winter because water temperature is lower and the days are shorter, although blooms of dinoflagellates *Chatoceras, Aserionella,* and diatoms *Nitzschia* and *Thalassiosira*, can occur. Diatom blooms may trigger an increase in copepods and then fish larvae. In spring, all plankton numbers reach a peak and fish eggs and larvae are most abundant. Concentrations decline through summer into autumn and diatoms are particularly low but several zooplankton including *Ceratium, Peridinium,* and *Noctiluca,* are very common. One of these, *Noctiluca scintillans* is bioluminescent. When disturbed by waves, boats, or waders at night each alga fires a flash of blue light. When only small numbers are present this may not be noticed but when the algae is so abundant that it can be seen as a pink bloom during the day, the night time effect along the shore can be very impressive.

Noctiluca was only occasionally seen north of Newcastle in the 1980s but has become increasingly common from Sydney to Tasmania since 1993. The reason for this is not clear but possibly relates to an increase in diatoms that may have flourished as a result of the southward extension of the East Australian Current.

During El Niño years (drought years on the land) strong offshore winds induce upwelling of deep cold water along the coast. This water is rich in nutrients and the upwelling causes phytoplankton blooms, some of which can be so dense that billions of individual plankton actually colour the sea, green, yellow, pink, red, or even purple. Abundant plankton generally means an increased fish catch, but some plankton blooms can be toxic to fish.

Biological soup or sewage?

Often on a hot day, when the low tide starts to turn and the incoming water creeps into pools, a brown scum can appear on the water surface, drifting wherever the wind pushes it. Many observers immediately seek a land-based and possibly human explanation for this scum and erroneously think of sewage.

The scum exists because the surface of any water body is a monolayer in which the water molecules are aligned to one another and have the ability to attract and hold organic matter and small particles. Part of the scum is plankton and the rest is derived from by-products of life such as waxes, oils and proteins.

Life and death are two parts of the energy cycle. Organisms are born, leaving behind empty eggshells. Shrimps moult, casting adrift protein-rich empty skins. Molluscs move over their own mucus film leaving behind a protein-rich slime. Our scum has started. As the pendulum of life swings many other products are added to this brown film. The daily waste from sea-urchins is added in the form of small green pellets, together with the waste of millions of other rock pool inhabitants: be it from fish, crabs, bristle worms, or sea anemones. Nothing is lost in this food chain of life. The waste of one organism is the staff of life to something else. Sloughed-off dead cells from algae or fish or coelenterate worm, bird droppings, remains of scavenged barnacles, unused gametes left over after the reproductive efforts of bivalve molluscs, anemone or brittle star; all go to strengthen that scum, to enrich it. Finally the scum becomes a biologically rich, extremely nutritious, soup. And that rich broth is a nourishing meal for many other organisms, recycling nutrients via this conveyor belt endlessly back into the system.

Life is death is life is death is life – forever.

The filter feeders are an essential link in the chain of life connecting the most abundant food resources in the ocean such as bacteria and phytoplankton with the higher organisms. Scum is fine if you are a filter feeder living in a pool where all the food resources concentrate, but how do others get on when they have to extract their microscopic tucker from the very dilute suspension of ordinary seawater? The answer of course is that they have to deal with a lot of water and that involves some form of pumping. Even the smallest cunjevoi can pass more than a litre of water in an hour and a study in Victoria's Port Phillip

Bay estimated that the entire volume of the bay (about 25 km^3) passed through the bodies of filter feeding organisms every week.

Catching the food is the next trick and that might involve sieving the particles through cilia, across the collars of sponges or trapping them in a net of mucus as in the polychaetes and tunicates. These mechanisms can trap particles (and organisms such as bacteria) that are as small as 0.1 microns across.

Brown scum on a warm rock pool at Long Reef, NSW.

Green, brown, red or blue?

The seaweeds, or algae as they are more properly called, are to the marine world what flowering plants are to the terrestrial world, but there are some big differences.

All plants photosynthesise, getting energy from the sun. Flowering plants generally have roots, which anchor them to the ground and supply them with nutrients and water. Pull a plant out of the ground and it dies. Most plants have a vascular system circulating moisture and nutrients in the sap to trunk, leaves, and fruits. Marine algae are not vascular. They have no roots and anchor themselves to the substrate with holdfasts. They synthesise energy from the sun, but absorb nutrients from the water through their surface. Algae might be broken off rocks by big seas, but as long as the plants remain bathed in sunlight and water they remain alive and healthy. What they don't like is to sink beyond sunlight, or to be stranded on a beach where they dry out.

The holdfast of *Ecklonia radiata* attached the algae to a large boulder, but not so large that it could not be moved by waves. Boulders like this are very effective in abrading shore platform surfaces when on the move.

Flowering plants have a flower, which serves a reproductive role. Algae don't have flowers, and their reproductive strategies are complex and varied. In fact there are very few flowering plants in the marine world, eelgrass *Zostera capricorni* and *Z. muelleri*, and *Posidonia* spp. being local exceptions. Both *Zostera* spp. are present in New Zealand waters but *Z. muelleri* is limited to the colder parts around Victoria and Tasmania. The distribution of both *Zostera* species is diminishing in most estuaries and bays along the East coast as the genera do not handle high nutrient loads, or disturbance by boat anchors very well.

There are three main types of algae; green, brown, and red. Most algae contain chloroplasts (green pigmented cells) but they also contain chromoplasts, organelles that have other pigments that can absorb the available solar energy. The pigment colours are complimentary to the wavelength of the available light. So the pigments of green algae absorb red and blue light near the surface. Brown algae with green and red pigments live a little deeper and red algae absorb the last available green light in the deepest water. Colour names are not always reflected in the actual colour of the plant. Thus 'blue green algae' are often black, and red algae can be shades of dull green, olive green, brown, or purple, as well as pink and red.

Underwater view of *Hormosira banksii* (foreground) and *Zostera capricorni* in a pool on Long Reef, NSW.

Codium fragile a local member of a very widespread family.

Common green algae on shore platforms include southern sea lettuce *Ulva australis*, a bright green seasonal alga found on the rock flats in the more open, wave-washed areas. Dark green dead man's fingers *Codium fragile*, and the southern codium *C. australicum* occupy pools. Baitweed *Enteromorpha* spp., is found at a higher level in the splash zone, often near freshwater seepages, and it seems to be an indicator of higher nutrient levels.

Brown algae are more common in slightly deeper water and are represented by common kelp *Ecklonia radiata* that is often washed up in large quantities on the beach after storms, and Neptune's necklace *Hormosira banksii* that carpets the big rock flats in summer. Each frond is made up of many individual spheres – hence the description 'necklace'. Like many algae, this species is seasonal, and towards the end of summer drifts of dead fronds carpet the tide lines. *H. banksii* reproduces asexually from broken parts and sexually from attached plants. The trigger for egg and sperm release seems to be cold, dry conditions in winter.

Padina sp. a tropical species that is near its southern limit in NSW.

Colpomenia sinuosa, a globe algae that is found right around Australia but only extends to about 1 m below low tide.

One of several genera of red algae encrusting rock surfaces below mid-tide levels.

Iridescent algae
Champia compressa

Thin-leaf fan weed *Padina sp.*, is an unusual form of brown alga with tough (partly calcified), concentrically banded fronds found during summer in the rock pools and at the low tide line where leathery looking globe algae *Colpomenia sinuosa* is also found.

Red algae are mostly found in the deepest water and they include some surprising variations to the normal, soft, pliable model that fits most algae. These include the encrusting red algae that coat the rock surfaces at the low tide limit in high-energy zones with a pink film hardened by calcium carbonate. These genera do not even look like plants. Differentiating the species is extremely difficult without expert knowledge and a microscope.

The tufted coralline algae *Corallina officinalis* and other coralline algae such as *Amphiroa anceps* are also found here. These species have a hard, calcium carbonate structure encasing the fronds, which one would think would make them very liable to breakage in strong wave action. However, they have overcome this by branching continuously with an 'elbow' at each joint that allows flexibility. They may be distinguished from one another as *C. officinalis* branches into three and *A. anceps* branches into two, every time a frond divides. After death coralline algae breaks up and becomes part of the sand.

Southern laver *Porphyra columbina* is a thin purple-red sheet that is common in winter. Relatives of this species are harvested elsewhere in the world as *nori* and *laver* and used to wrap *sushi* (Japan) or *laverbread* (Wales).

Another red alga *Champia compressa*, is a conspicuous bright fluorescent blue or green that stands out in any tide pool and at the low tide level of the shore platform. If you see it out of the water the colour vanishes because it is created by light interference and is not based on colour pigments.

Oddities of reproduction

Reproduction is all about passing on genes from one generation to the next and there are two ways to do it, asexually or sexually. In asexual reproduction (budding) there is little recombination of the genetic material and offspring are nearly identical to the parents. In evolutionary terms this is a very conservative route that is mainly practised by the simplest forms of life such as viruses and bacteria but it can extend to organisms well up the evolutionary ladder such as some of the coelenterates (corals and anemones).

Sexual reproduction on the other hand shuffles the genetic material between generations. This results in more variety in the offspring and provides opportunity for evolution of new species. Sexual reproduction requires a male and female of the same species to get together but that's more than a little difficult if you happen to be attached to the rocks of a shore platform. Consequently sessile (attached) organisms have developed strategies that reduce the business to simply enabling the eggs and sperm to get together in the water. Here nature takes its course and reproduction becomes a great game of chance: the chance meeting of a particular sperm and egg, the chance of not being consumed by something, the chance of lodging in the right place to grow into an adult, and the chance of surviving long enough to mature and drive the cycle again. When the odds are against you so heavily one answer is to flood the market with enormous numbers and perhaps to use a few special tricks like releasing pheromones (chemical attractants) to entice potential partners or juveniles to the right locality.

On the shore platform many organisms have male and female functions in the one individual, either at the same time (hermaphrodites) or sequentially (sex change). We find asexual reproduction in some of the higher groups but even amongst these we expect to see a sexual stage somewhere in the life cycle, and we are often faced with puzzles about why particular species have extended population ranges with little variety in form, and yet they do not seem to have any means of exchanging genetic material across long distances.

For example, consider two common intertidal sea-stars, the eight-arm sea-star *Meridiastra calcar* and the rockpool star *Parvulastra exigua*. Individuals of the eight-armed star are either male or female. They breed normally and the larvae disperse in the plankton. Individuals of the rockpool star, however, are hermaphrodites. Larval development to the adult stage takes place in an egg mass attached to the rocks close to the parent and there is no obvious means of long distance travel. The species has a wide geographic range and genetic studies reveal that there are like clusters on different headlands but apparently isolation is not so tight that these groups have become separate species. Somehow enough get around to maintain a single interbreeding population, but we have no idea how they do it!

Clonemate group of *Actinia tenebrosa*, tentacles retracted.

An even stranger case is seen with the waratah anemone *Actinia tenebrosa.* This species produces juveniles asexually inside their bodies and in late spring a cluster of clones (genetically identical offspring) gather around the parents. The adult even has special stinging cells below the feeding tentacles that are used to attack non-clonemate juveniles. So all the adults on large areas of the shore platform are identical. But the range of *A. tenebrosa* is right around the southern coast of Australia and across to New Zealand, so there also

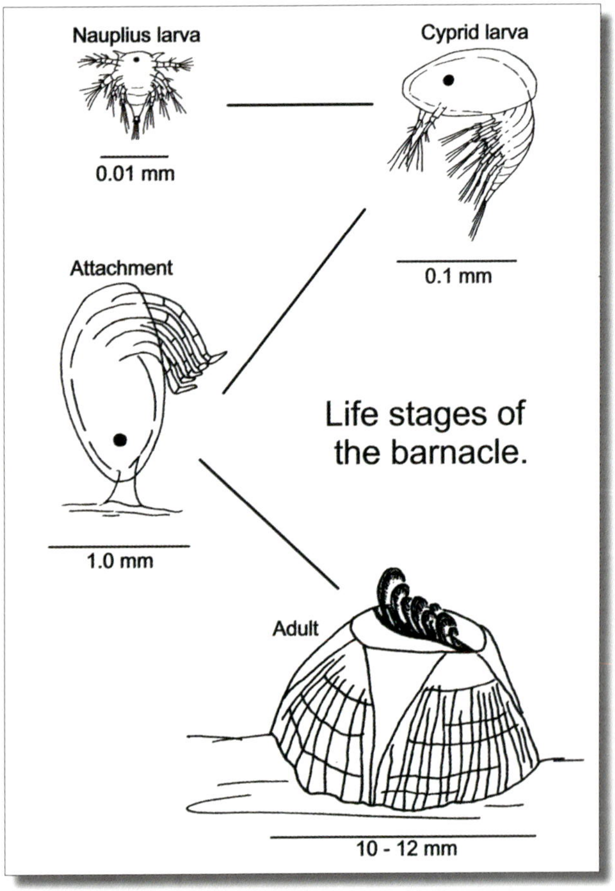

Schematic life cycle of a barnacle.

has to be a sexual reproduction stage to enable genetic exchange across these distances, but we have not yet identified it.

Incidentally, *A. tenebrosa* is not as sedentary as it seems. It can actually move around on its muscular base.

The hermaphrodite list includes; nudibranchs, the rose-coloured barnacle *Tesseropora rosea,* and the rosette barnacle *Tetraclitella purpurascens.* Like most hermaphrodites the barnacles do not self fertilise but live in dense colonies and cross-fertilize one another using a very long penis. Changing water temperature and day length apparently trigger breeding. The developing eggs are retained in the body for several weeks up to the first larval stage, the *nauplius*. Then they are released when the phytoplankton on which they feed is abundant. They spend a few months in the water as part of the zooplankton moulting to the *cypris* stage. In this form they do not eat and can survive about 12 days, then on receiving a chemical signal in the water thcy settle on the rock attaching by their antennae and secrete adult plates within 12 hours.

Two of the great 19th Century zoologists, Louis Agassiz and Thomas Huxley once described barnacles as, 'a shrimp like animal standing on its head in a limestone house and kicking food into its mouth'. Yes, that is same Huxley aka 'Darwin's Bulldog' who so ably defended Darwin's theory on the origin of species in 1860 and who had previously spent almost four years as a zoologist sailing the waters of Australia and New Guinea in *HMS Rattlesnake.* In that time he studied and later published major works on the Cnidarians and Tunicates, and whilst in Sydney found the time to court his future wife, Henrietta Heathorn. When we spend time on the shore platforms we do indeed 'walk in the footsteps of giants'.

Many marine molluscs exhibit sex reversal with age. Young limpets for example are all male and as they get older become female. The same is true for rock oysters. This is better considered as a form of sequential hermaphroditism and makes an interesting contrast with the terrestrial gastropods (garden slugs and snails) that remain hermaphrodites for life. A good case has been made for slugs and snails that being hermaphrodites optimizes the chances of reproductive success in species that have low population density and/or limited ability to movc around. For these animals every meeting between adults is an opportunity for reproduction. But sequential hermaphroditism is more puzzling.

Eggs

In sexually reproducing organisms, biologist distinguish direct development from indirect development. Direct development is when the offspring have similar form to the parents and develop through steady growth, for example; some fish, all birds, and mammals. Indirect development involves offspring that are quite unlike the parents and develop though a series of larval stages and metamorphosis, for example most invertebrates. What are the costs and benefits of these strategies?

The eggs of directly developing organisms are larger and fewer, and most parents spend time and energy in nurturing the very young. On maturity they are ready to go and independent. The downside of all that parental effort is that in adverse conditions both parents and young may succumb to disaster, and population recovery can be slow. Organisms using the indirect strategy have considerably larger numbers of smaller eggs that require little or no care. At each growth stage larval offspring may utilize different foods or habitats and there is less competition between adults and young. Saturating the environment with large numbers of offspring provides a larger food resource to predators, gives greater assurance of species survival, gives more opportunity for widespread dispersion, but increases the chances of loss of each

Stranded egg of a Port Jackson shark.

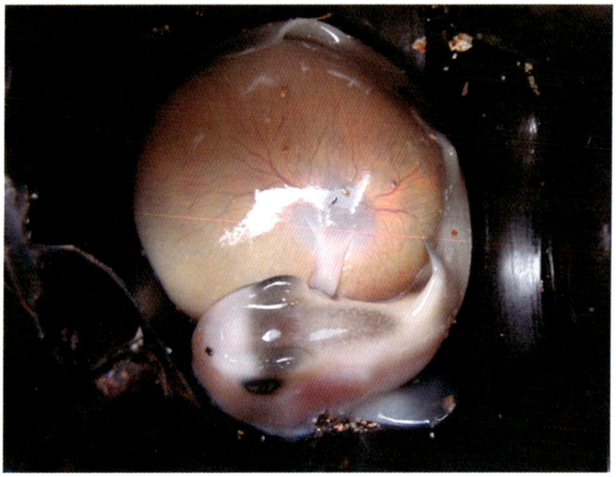

Embryo of a Port Jackson shark attached to the egg yolk.

individual. But even in adverse conditions there are always a few who survive to maturity and they can rebuild the population quickly.

For organisms using the strategy of direct development for reproduction there are many ways of packaging the eggs to improve their chances of survival. One of the neatest examples is practiced by the common Port Jackson Shark *Heterodontus portusjacksoni*. These sharks grow to just under one metre in length and are found in bottom waters down to about 100m all around southern Australia. They are not harmful to humans as their teeth are developed to crush and grind not cut and tear. Favoured foods are molluscs, crustaceans, fish, and long-spine sea-urchins *Centrostephanus rodgersii*. They breed in the winter and are oviparous, that is, they lay eggs like birds or turtles.

Young adult *H. portusjacksoni*. Photo by Dave Thomas.

These sharks make an elaborate egg case, 12 to 15 centimetres long containing one large-yoked egg. The case is a dark brown horny material with a curious corkscrew flange around it and a tangle of wiry filaments extending from one end. The sharks wedge the cases into rock crevices on the seabed or the filaments are entangled in the holdfast of kelp and there they remain for the 10 to 12 months required for the embryos to absorb the egg yolk and break out of the case as a juvenile shark. You can often find hatched eggs washed up on the beach.

Good packaging of eggs is intended to protect the developing young until they are ready to brave the outside world. The phylum Mollusca has by far the most species producing capsules, so the following remarks are directed to that group.

The four centimetre half-dome egg case of the common Sydney rock whelk *Cabestana spengleri* consists of a large number of stalked capsules filled with many developing larvae. You will find these common shells and egg masses at low tide level amongst the cunjevoi, the gonads of which are its main food. Unlike other whelks *C. spengleri* does not drill holes to reach the food but passes its proboscis down one of the siphons of the cunjevoi and attacks directly.

The common cone shell *Conus anenome* produces a neat row of individually deposited capsules, again containing more than one individual. The adult of this species may be found 'sitting on' and protecting the capsules until the young emerge. Protection of eggs is rare. Cowries also do it, resting on their cluster of soft globose capsules.

Some egg masses are easy to identify. In Winter and Spring the cartrut shell *Dicathais orbita*, lays clusters of capsules that look a bit like a sunflower head without the petals or pieces of honeycomb. When first laid the eggs are pale yellow flattened cylinders 5 to 8mm long. They gradually change to purple just before hatching. Bleached empty egg masses often wash up on the beach to puzzle observant visitors. The colour change is caused by chemical changes in the egg that end in the formation of Tyrian purple dye – the same material extracted from marine gastropods in ancient times and used to dye Imperial Roman togas. The chemical precursors to the dye have strong anti-bacterial properties and apparently protect the developing embryo from infection. The dye itself has little protective value. *D. orbita* is a common predator on many shore platforms around southern Australia, keeping barnacles and mussels in check and studies in Western Australia have shown that some individuals can live to 20 years which is good going for a mollusc.

Egg mass of *Dicathais orbita*, note the mauve one in the centre.

Unknown purple eggs found on the underside of a rock at the low tide level. Some have either been hatched or eaten leaving circular scars.

Another egg structure occasionally seen is that used by a member of the carnivorous family Naticidae. The relatively common smallish sand snail *Natica sagittata* produces a very neat 'collar' some two centimetres in diameter, in which individual eggs are laid within a matrix of cemented sand grains (see heading p 68). This species has a relative, the sand snail *Conuber sordidum (Polinices conicus)*, found on the estuarine mud and sand flats. This produces a large (six centimetres) jelly sausage consisting of a clear jelly matrix in which are embedded thousands of barely visible eggs.

Sea slug egg ribbons may be brightly coloured and are attached to the rocks as a coiled ruff.

Some of the very common intertidal and supra-tidal gastropods of the *Littorinidae* produce a small, half-spherical clear fluid-filled disc in which one egg is deposited. These eggs are dropped into the water and become pelagic, a neat way of preventing desiccation if you choose to live so high on the shore.

When the tide is low and a warm pool is open to the sun, a procession of Sydney sea hares *Aplysia sydneyensis* can be seen moving along with reproduction in mind. They even use tidal swimming pools. After mating long, thin strings of orange or yellow eggs, looking rather like vermicelli are laid in a compact tangled mass. A record for one individual was to lay a 600m long egg string in four months. The eggs are left to hatch at the whim of wind, tide, and temperature.

Like sea hares many other organisms gather together during the breeding season. A full moon, an unusual tide, or a temperature change, can trigger mass spawning. We know surprisingly little about these triggers. One recent study tried to establish the causes of timing in 18 species of gastropods by regular survey of the presence or absence of egg masses over two years. Most species did have peak spawning periods, usually in Spring and Summer, but none of the triggers claimed in the literature were found to be valid. The results suggested two explanations: either the waters of south-eastern Australia are nutrient-poor in comparison to other parts of the world where the classic patterns have been described and perhaps food availability is not a critical factor, or the whole business has much more to do with adult physiology than we have appreciated.

Sponges

Sponges are the simplest of all multi-celled organisms. They are placed in the animal phylum Porifera, because the body wall has only two layers of cells. The wall is perforated by numerous holes through which water is moved by lots of tiny flagella allowing the sponge cells to filter food from the water and to dispose of wastes. There is no circulatory or nervous system, no specialised digestive system, no mouth or anus, in short, almost nothing but sponge. Each cell is able to perform all the essential functions and you could chop a sponge into bits and all could survive.

The internal structure of the sponge includes an intricate skeletal framework that may be familiar to you as the once common bath sponge that was harvested from Mediterranean waters. In recent years this has been replaced by brightly coloured synthetics that carry the same name. Natural sponge skeletons are composed of interlocking spicules of calcium carbonate, horny organic matter, or more rarely beautiful opaline silica. Considering that silica is almost insoluble in water and has such a low concentration in seawater, sponges with silica spicules must circulate an enormous volume of water to get enough to grow unless perhaps they obtain it indirectly by ingesting other siliceous organisms such as diatoms. The living sponge bears little resemblance to the dead sponge skeleton washed on to the shore and in life the body of the sponge commonly provides habitat for dozens of other organisms that shelter within it.

The shape of the animal, (yes it is an animal, not a plant, as it has no chlorophyll), varies a lot but in shallow waters its colour and surface texture are consistent and the common shore platform species can be identified fairly easily. The purple encrusting sponge *Haliclona* sp. is common in rock pools, especially those exposed to full sunlight. The colour is

The purple encrusting sponge, *Haliclona* sp.

consistent but the shape reflects the substrate. They have siliceous skeletons and the body is covered in crater-like holes through which they pass water for feeding and respiration.

A sponge, probably *Thorecta* sp., washed up on the beach as flotsam.

As the name suggests the finger sponge or cream coloured sponge *Holopsamma laminaefavosa* looks exactly like a bunch of pale orange to brown fingers with a honeycomb surface. It lives in the sub-tidal zone and dead specimens are common on the beach after a storm.

The rose sponge *Dendrilla rosea* occurs in slightly deeper water and can sometimes be seen transplanted onto decorator crabs or encrusting larger shellfish.

The liver sponge *Chondrilla australiensis* also encrusts stones as a 10mm layer up to 250 or 300mm across at low tide level. This sponge often has bivalves or small anemones growing within it.

Another dead sponge washed onto the beach, probably *Holopsamma* sp.

Ectoprocts

Commonly called lace corals, sea mosses, sea mats, and moss animals, the bryozoa are members of the phylum Ectoprocta. They are tiny, colonial organisms that have a long fossil history and an unusual development pattern. Individuals within the colony are referred to as zooids and they are enclosed within a protective case (test) one to two millimetres long composed of calcium carbonate or chitin, the same material as insect exoskeletons.

Their form ranges from flexible, horny mats, to rigid calcareous lace corals *Triphyllozoon moniliferum*, to branched and projecting 'plants' much unloved by sailors who have a never-ending battle with hull fouling.

Bryozoans are filter feeders but we know little about their diet and assume that it consists of very small organisms such as bacteria, flagellates, and diatoms. But there is a possibility that some may gain nutrients directly from the living seaweed they colonise. Experiments using radioactively labelled carbon in kelp have traced nutrient flow into the bryozoan living on the kelp. Perhaps this

Lace coral *Triphyllozoon moniliferum*

Colony of *Membranipora membranacea* on a kelp blade.

parasitic(?) relationship is more common than we realise. Seaweeds, however, are not without some defence. Species at risk from suffocation by blankets of bryozoa produce chemicals that repel these and other epiphytes.

About 5,000 bryozoan species exist around the world and 210 have been recorded from Sydney Harbour. Due to their ability to encrust floating objects such as pumice, coconuts, and bottles, some species are widely distributed and new arrivals are frequent on our shore.

Those with spiny skin

Starfish do not swim and are only distantly related to fish. They are better named sea-stars and are part of a diverse group of organisms in the Phylum Echinodermata. The phylum also includes brittle-stars, sea-urchins, sea-lilies or feather-stars, and sea-cucumbers or bêche-de-mer. At Long Reef you may occasionally find a large dark brown or black sea cucumber either *Australostichopus mollis* (*Stichopus mollis*) or *Holothuria leucospilota* crawling slowly on the bottom of a gutter at low tide. *H. leucospilota* has a tropical distribution and has only been seen at Long Reef in recent years. We wonder if its presence is an indication of warmer conditions and climate change.

Holothuria leucospilota, if roughly handled they will release a mess of sticky white threads that may be toxic.

What do all of these different looking organisms have in common? The phylum is exclusively marine, all members have radial symmetry of the body parts, and an amazing ability to regenerate damaged parts even to replacing arms. They also have a unique plumbing network (vascular system) that pressurizes the body structure and in most species drives hundreds of tiny tube feet out from the body to move them around, hang on to the substrate, find and capture food, and enable respiration.

In sea-urchins and most sea-stars the mouth is on the under surface (the oral or ventral surface) at the centre of the body and the anus is on the upper (aboral or dorsal) surface. Turn over an echinoderm and you will see the tube feet waving around seeking something to hold in order to turn the animal back up the right way.

A sea-star can extend its stomach outside its body through the mouth and envelop its food, the food being shellfish, particularly bivalves. But how can a weak little sea-star open a bivalve? With persistence is the answer! By wrapping its arms around the shell and exerting a small but steady pull with the tube feet it eventually overpowers the stronger muscles that close the shellfish as they tire. As soon as the shell gapes, in goes the stomach and dinner is served.

Meridiastra calcar oral (ventral) and aboral (dorsal) surfaces. Colour can vary enormously in this group.

M. gunnii, a six-arm sea-star with rather duller colours.

Coscinasterias muricata also found in tidal waters of NZ.

Common sea-stars include the eight-arm sea-star *Meridiastra calcar*; the small green rock pool star *Parvulastra exigua* with its five arms reduced almost to knobs; the less common six-arm sea-star *Meridiastra gunnii* (*Pateriella gunnii*); and the large eleven-arm sea-star *Coscinasterias muricata* (*C. calamaria*). Mind you, this last one can have anything from 7 to 14 arms, with several of different lengths and in the process of being regenerated. The 'lost' arms may have been taken by a predator, or may indicate that the animal has recently divided asexually.

A marine version of peanut brittle? A peanut worm *Phascolosoma noduliferum* and a brittle-star *Ophiothrix spongicola* found under the same boulder.

Most of the short armed stars found in rock pools and crevices show a wide range of broken colour patterns in shades of grey, green, blue, orange, mauve, and red. These patterns make them quite cryptic and presumably protect them from predatory fish. The pattern on these organisms can be described as pixelation, a concept that is now familiar to us through computer screens. It has only recently been appreciated by military artists in providing improved camouflage over the more traditional dazzle patterns made with random blobs and streaks. This reminds us of a little known fact that during WWII Professor W.J. Dakin, the Sydney zoologist and senior author of our most popular text *Australian Seashores*, was seconded by the Commonwealth Government because of his special knowledge to become the Technical Director of Camouflage in a civilian unit at George's Heights, Sydney.

Brittle-stars are found under rocks on the shore platform. They belong in another class of the echinoderms, the Ophiuroidea. Their five arms are much longer than the sea-stars and they move around in sinuous undulations that are very like a snake's movements. The arms break easily if the brittle-star is handled, so best leave them alone. In brittle-stars the calcareous plates that make up the skeletal framework of both sea-stars and brittle-stars are more obvious and each arm is a series of hard rings joined by muscle. They do have tube feet but these are not as important as in the sea-stars and less obvious. The mouth is on the underside as usual, and is equipped with small teeth and it also serves as the anus. The species filters bottom sediments to recover food. Common examples are the southern serpent star *Ophiomyxa australis,* the large spiny-armed brittle-star *Ophiothrix spongicola,* and the brown-banded brittle-star *Ophionereis schayeri.*

On worms

It is almost impossible to write briefly on 'worms'. It is just about as difficult to know what is meant by the word. This problem is a legacy from Linneaus who 300 years ago dumped all the simple animals that he had difficulty sorting into one category '*Vermes*' or worms. We still find that many totally different organisms ranging in length from a few millimetres to more than a metre go under this common name. Many worms belong in the following six unrelated marine phyla:

Phylum Annelida

The most common of all the worms are annelids or segmented worms. They are divided into four classes that include earthworms and leeches, with the leeches being a common but little known parasite on higher organisms in the marine world. The class polychaeta stand out, those that are free-living and mobile and those that construct tubes of carbonate, cemented sand, or shell fragments.

Some eat their way through sediment, some eat algae, and others are predators. Fish, crabs, predatory snails, wading birds and fishermen seeking bait all hunt them. Each body segment carries a set of prominent external bristles, called setae, which are used for movement, to help anchor the worm, and for other purposes. The group includes the fire worm *Eurythoe complanata*, one of the few organisms to be avoided in the temperate marine world. Salmon-pink to red in colour, this worm is seldom more than four centimetres long but can inflict intense pain if you brush a finger against its setae. Be careful when probing through rubble and gravel under stones, its favourite habitat.

One of the largest worms to be found on a rocky shore is the quite beautiful and iridescent biting worm *Leodice laticeps,* which can reach over one metre in length and be about one centimetre thick. It lurks under boulders in the lower intertidal zone. Relatives include the giant beach worms, one species of which can attain a length of three metres. Beach worms are known to fishermen as great bait and referred to as stumpies or kingworms, redheads, bronzies, whiteheads, or slimies. There are several species of *Australonuphis* involved in which the sexes and the juveniles have different form and size. These organisms glide through the sand in the intertidal zone and are omnivorous scavengers and predators of shellfish and other organisms.

The feather-duster worm *Sabellastarte australiensis* (*S. indica, S. fusca*) is a polychaete that constructs a permanent home in a tube, emerging only to feed when things are quiet. It whips its feeding tentacles back into the tube at the slightest disturbance. Similarly the Sydney coral worm *Galeolaria gemineoa* occupies a limy constructed tube, and numbers can be so dense that it is an indicator species for the mid-tide level just above the cungevoi. Both *S. australiensis* and *G. gemineoa* have numerous tentacles covered in fine beating hairs that filter food particles from the water. *Spirorbis* spp., are in a closely related group and form many individual and very small white coiled tubes, adhering mostly to the underside of rocks in the intertidal zone. Occasionally they will be found on the blades of sea grasses and on the hard surfaces of free-living animals such as crabs and molluscs.

Typical density of Spirorbid worms on the underside of a boulder. Each worm shell is less than 3mm in diameter, species unknown.

A tube about 10mm in diameter constructed of cemented platy shell fragments is easy to find under boulders on the shore platform. This is the home of another polychaete, the Sydney parchment worm *Diopatra dentata*.

A flatworm probably *Callioplana marginata*

Phylum Platyhelminthes

This phylum covers a very large group of worms that includes the class Turbellaria or flatworms. Most are found in tropical waters, but local examples include the orange margined flatworm *Callioplana marginata* found under stones and the southern flatworm *Notoplana australis*. Some species can be confused with certain nudibranchs, which, among other phyla, they mimic superbly, but closer inspection shows that their common name is very apt: they are very flat and thin. They are carnivorous, with the mouth under the body and near the front. They swim with an undulating motion.

Phylum Sipuncula

The nodular peanut worm, *Phascolosoma noduliferum* is a small brown unsegmented, cylindrical worm less than three centimetres long. It lives in rubble under rocks, or entwined in Sydney coral worm tubes. Peanut worms feed on detrital matter by extending a sort of proboscis that holds the mouth at its tip. This organ is normally held within the body but can be extended to more than quarter of the length of the animal.

Phylum Nemertea

Their common names are ribbon worms or proboscis worms. They also have an extensible proboscis that is used to capture prey, which is very variable; molluscs, crustaceans, or other worms. Most are generally quite small, unsegmented and quite common. They are actually quite important, though they are seldom noticed. One exceptional example is the brilliant orange ribbon worm *Gorgonorhynchus repens* found under stones often coiled like a knot.

Brown nemertean worm trying to tie itself in knots, species unknown.

Phylum Nematoda

A shallow dish of sand from a rock pool or beach will doubtless show at 10 to 20 times magnification, a small, actively wriggling colourless worm, or two, or several, or many. These are nematodes, largely unseen and unnoticed but incredibly important, recyclers of organic matter and nutrients. They are featureless and unsegmented, with the head at one end and anus at the other. Most are free living although some are parasitic.

Phylum Mollusca, Class Bivalvia

This brief review of 'worms' would not be complete without mentioning the southern shipworm, or teredo *Bankia australis* (and other species), which is actually a boring bivalve that drills large holes in wood. Worm-like, but hardly a worm! Oh Linnaeus, what did you start by not completing this part of your classification?

Limy tubes of *Teredo* sp., in driftwood.

Home range and locking down

How far does the average snail wander on the shore platform? How does it find its way around and how does it know where home is? Such questions have been asked by numerous zoologists, ecologists, and people who study animal behaviour (ethologists), and the answers are intriguing.

Moving at a snail's pace is a common expression and it really is pretty slow. Snails have been timed on the shore platform moving at 7 to 12 metres per hour. This figure tallies with other observations recording total movements of between 20 and 40 metres per day (24 hours). And if you happen to see a gastropod that is exceeding the speed limit you are probably looking at an empty shell that has been taken over by a hermit crab such as the boxer hermit crab *Paguristes pugil*, so named for its pugilistic behaviour! Hermit crabs belong to two families Diogenidae and Paguridae. The larger ones are usually Diogenids. What a lovely name! Diogenes was an ancient Greek philosopher, who allegedly made his home in a barrel and lived on a diet of onions. Look closely in any rock pool and you will be quite surprised at how many old snail shells suddenly move. Pick them up with care and instead of a slimy foot or closing operculum you may see arthropod claws and legs withdrawing. Hermit crabs only have a hard exoskeleton on the thorax, the coiled abdomen is soft. They squeeze the abdomen into the smaller spaces in their recycled home and fold the claws up in the mouth of the shell where they act and look like an operculum. Every time the animal goes through a moult cycle they have to find a larger shell and try it for size.

In living snails several common patterns of movement are recognised. Most limpets including *Siphonaria* spp., and often the variegated limpet *Cellana tramoserica*, return precisely to their home scar. Others species return to a general location or the same tidal level but not exactly the same spot. Snails that aggregate in the shelter of rocks such as the blue periwinkle *Austrolittorina unifasciata* commonly return to the same cluster but there are species whose feeding movements appear to be totally random and they can end up in a new location after every trip.

It is difficult to describe specific patterns of behaviour as longer-term studies show that behaviour can vary over time. One study of *Cellana tramoserica* showed that some animals take several days to complete a circuit, some move to a new

All of the small gastropod shells on the underside of this boulder are occupied by hermit crabs.

position and then begin homing, and some just move about or perhaps home for a few days and then move on. What drives the behaviour? To answer that, we have to speculate. It has been suggested that homing is a simple mechanism for effective sharing of the food resource on the shore platform but that of course is only a partial answer. It does receive support from animal distribution studies that measure distances to the nearest neighbour, which show that there is some process of active dispersion going on in local populations. The grazers certainly seem to have stronger homing tendencies than the predators and the idea of a home range or a regularly grazed surface makes sense.

The question of how they know where home is has not been fully answered. Some animals backtrack on their mucus trail but others complete a one-way circuit and appear to know where they are at any time. Can they follow old trails? Perhaps, but trails probably decay within one tide cycle and this would not be a very reliable guide to home location. The specialists suggest that homing is learned and point to the fact that the average gastropod has sufficient neurons (about 20,000) to service basic memory and learning.

Then there are the gastropods that take a ride on other species and don't move around very much at all themselves. The small limpet *Patelloida mufria* is mostly found on the shells of other gastropods, especially the ribbed winkle *Austrocochlea constricta*. It much prefers winkles that are living in tidal pools. Should the winkle move out of the pool the limpet abandons it and seeks a new host, as it is apparently susceptible to drying out. What do the parties get out of this commensal relationship? The host winkles get clean shells as the limpets graze their surface, and the limpets get protection from predatory mulberry snails like *Morula marginalba* because their hosts are mobile.

Opercula

It is fascinating how you can get carried away and led up many paths while you investigate the origin and meaning of a particular word. Open most text books, turn to the section on gastropod molluscs, and there it is, the operculum, a calcareous door attached to the end of the foot and pulled in last to seal the animal from the outside world. But open a good dictionary and the answer is different. Not all opercula are calcareous, not all gastropods have them, barnacles also have them, so do certain tubeworms, so do fish, and the path branches even more when we learn that they are also present in some birds where they serve as a cover for the ears and nostrils. Eucalypts have them as bud covers protecting the flowers - the pixie cap, and even fungi have an operculum!

The word is Latin and simply means 'a cover or lid'. But to most people an operculum is the solid door, or more correctly lid, used to close the aperture of many gastropods. In the terrestrial world some land snails form a temporary and sometimes quite solid plug of mucus that acts as an efficient seal, except that it is water-soluble and is lost when it rains. This is called an epiphragm.

Turbo torquatus with a ribbed operculum and *T. militaris* with a smooth operculum.

Some gastropods have small, remnant horny opercula that probably don't provide that much protection at all. Other molluscs, particularly Turbinidae and related families carry a very solid, sometimes heavily sculptured and sometimes highly coloured operculum, known as a 'cat's eye'. Some of these may be familiar to you as jewellery.

Tent shell *Astralium tentoriformis* showing the operculum and a small chiton on the underside.

All of the opercula composed of shell are tough and durable and you will find battered examples on coarse sandy beaches on high-energy coasts and frequently in Aboriginal shell middens.

Octopus and relatives

Octopus, along with squid, cuttlefish, the pearly (emperor) nautilus *Nautilus pompilius* and a small spiral shell from the rams-horn squid *Spirula spirula* are all molluscs in the class Cephalopoda. How are they different from other molluscs?

Spirula spirula are commonly washed onto the beach but very rarely seen alive in local waters. The largest examples are about 30mm across.

Well, to begin with, they are all free-living predators that can 'swim' using a jet-propulsion technique. For protection most can squirt an ink-screen into the water behind which they disappear. *Nautilus pompilius* and all the blue-ringed octopus *Hapalochlaena* spp., are exceptions. It is uncertain whether the ink is only a visual screen or if it also acts as a chemical deterrent. Some species have the ability to rapidly change their body colour to match a background. This is done with specialised skin cells that contain brown, black, red, and orange pigments that can be flashed at will with colour changes taking less than a second. The blue-ringed octopus *Hapalochlaena* spp., are so named because when provoked they become bright yellow with blue markings, often chevrons, and in the species found near Darwin, vivid rings. When undisturbed the blue markings are much less prominent. One species can be found amongst rocky rubble or dense seaweed on most east coast platforms, although it is difficult to locate.

You need to be wary of any blue-ringed octopus as their bite can be fatal. Do not put your hand into places where you cannot see what is there and do not handle a blue-ringed octopus if you do find one.

Cephalopods have a complex nervous system and a large brain relative to body weight, larger on average than any animals except birds and mammals. The brain is linked to well-developed eyes that are also used as a sort of levelling device, in that the animal knows which way is up and the eye slit is always held horizontal.

Octopus and squid have no shell. In cuttlefish the shell is reduced to an internal structure or 'cuttlebone' that is often washed up on the beach. The common east coast species are the giant cuttlefish *Sepia apama* and the mourning cuttlefish *Sepia plangon*, both of which have the ability to totally change their colour patterns in a moment in response to external conditions. Both *Nautilus pompilius*, and *Spirula spirula* have chambered external shells.

Octopuses have eight arms, or are they feet? Squid and

Blue-ringed octopus *Hapalochlaena* sp., from Sydney. Photo by Jack Hannan.

Sepia apama at Manly NSW. Photo by Dave Thomas.

Cuttlebones stranded on the beach.

cuttlefish have ten. Of all these species you will only find living examples of the octopus on the shore platform, as the others live in deeper water although their remains are sometimes washed ashore.

Octopuses are generalists when it comes to feeding (although the blue-ringed octopus is nearly exclusively a crab feeder). The gloomy octopus *Octopus tetricus*, is found in gullies and crevices or under large stones on the shore platform. *O. tetricus* moves around in the shallows crawling on its arms, swimming, or jetting and it will catch what it can, from bivalves to crabs, from fish to cowries. As the tide falls they will often move from one rock crevice to another and they usually have a preferred den that is marked by a pile of crab and shellfish debris that are the remains of previous meals. *O. tetricus* can grow to about 80cm long but as with most cephalopods it only has a short life span. Females lay an egg mass in rock crevices and guard and clean these rubbery eggs during incubation.

The gloomy octopus *Octopus tetricus*, photo by Dave Thomas.

In evolutionary terms coiled shells and later straight shells of cephalopods are common fossils in rocks 444 to 251 million years old but only a few species survived the mass extinction event at the end of the Permian period. This makes *N. pompilius*, and *S. spirula* rather special but on the Australian coast we only ever see them as empty shells on the beach. *Spirula* is common on east coast beaches, *Nautilus* is extremely rare in the south. Both shells are quite unlike the coiled gastropods that they resemble, in that the shells are chambered with each chamber of the *Nautilus* having once been the home of the animal at a different stage of growth and then the empty chambers are used as a flotation device.

Sticky mysteries

Several species of barnacles are abundant on the rocks of the east coast shore platforms. The biggest are found in the most exposed locations where they crowd together and are only occasionally detached from the rocks in the roughest of storms. However, when an individual barnacle dies the shell will literally fall off the rock with the slightest disturbance.

And it is not just barnacles that stick so firmly whilst they are alive. What is the formula for the glue that anchors the bottom valve of an oyster to solid substrate, that prevents an ascidian from being dislodged from a rock, or that allows a bryozoan to stick to a floating object far out at sea, such as a log, a cuttlebone, or even a plastic bottle? Commercial adhesives are unsatisfactory in salt water, very few adhere to plastics, and to discover the formula for barnacle glue could be to unleash a new industrial product. Using clues from our limited knowledge of marine adhesives, one synthetic glue that can bind living bone slices has been prepared in the laboratory from simple components of gelatine plus a barnacle enzyme.

Scars left after barnacles have died and parted from the rock surface.

Base of *Austrobalanus imperator* showing the large surface area that barnacle 'cement' can adhere to.

Molecular biologists have identified unusual amino acids in these glues and we now know that they are high viscosity complex proteins secreted by the organism that fill the space between the organism and the rock. All are cross-linked molecules that resist pulling apart and they have the ability to exclude the water layer from the interface. We think that continued secretion of the glue is necessary to hold the organism in place, and that would explain why dead barnacles fall off.

Algae use glue too. Green and brown seaweed zoospores first attach to the rock with their flagella and then secrete 'glues' to bind themselves more permanently. Other seaweed zygotes have a thin layer of glue or mucus for initial adhesion and then develop rhizoids that eventually become the disc-like holdfast. Mussels secrete a two-part glue that is very like epoxy cement. It comes from a gland in the foot that can be extended some distance from the shell. Each strand or byssal thread takes about 10 minutes to set and then continues to harden and turn brown (tan) over a few days. Each thread is extremely strong but only about 25 microns in diameter. That is one quarter of the diameter of human hair.

The hairy mussel *Trichomya hirsuta* that attaches to the rock with byssal threads much like the threads on the shell.

Mobile animals such as snails, limpets, and sea-star all secrete a layer of mucus from their feet. This material bonds quickly and improves the animal's ability to hold on by suction. But it must also allow the animal to move when it needs to. There are two approaches. Some use viscoelastic mucus that behaves as both a liquid and an elastic solid depending on the applied stress, others involve another secretion that can chemically alter the mucus to allow quick release when the tide comes in.

In some species the adhesives may be toxic or distasteful to potential predators and thus the glue becomes a defence mechanism. How is that for lateral thinking?

Drifting barnacles

Whilst wandering along the shore you are very likely to find goose barnacles and bryozoa attached to pieces of wood, pumice, or other floating debris. Until recently we would have identified these simply as the long necked or the short necked barnacle and would have assumed that the object to which they are attached could have come from anywhere as these barnacles were thought to be globally distributed.

Stalked goose barnacles probably *Lepas australis* attached to a plastic rope and bucket.

The tale of the strangely named barnacles goes back through Charles Darwin to Linneaus and beyond to Aristotle (384 to 322 BCE). Linnaeus named two common forms of *Lepas,* one after ducks *L. anatifera*, and another after geese *L. anserifera* literally the duck-necked and goose-necked shellfish. Why would the father of zoological nomenclature create such a mixture of animal terms? The answer to that goes back to Aristotle and then to the middle-ages when people believed that these pelagic shellfish were actually an early stage in the development of the barnacle goose of the northern hemisphere *Branta leucopsis*. People thought that the barnacles were formed by spontaneous generation on timber floating in the sea. It was said that the young birds held on to the wood by their beaks (the shell plates) until the feathers (the barnacle feet) developed. They then detached from the wood and became adult birds. This was part of the theory of spontaneous generation for the origin of life formalised by Aristotle and it remained a basic belief in biology until finally shown to be wrong by Louis Pasteur in the 19th century. Would you have believed that a common barnacle could make all those links with the past?

Until recently we have been content to apply these universal names to drifting barnacles found in Australian waters, but the taxonomists have again revised the story and we now recognise the small goose barnacle *Lepas pectinata* in northern waters and the southern goose barnacle *L. australis* in cooler waters.

A block of pumice loaded with barnacles or other hitch-hikers is another story. Pumice is a frothy volcanic glass that is light enough to float. There are no active volcanoes in Australia but pumice often turns up on the coast. Because different eruptions produce pumice of different composition the sources can sometimes be traced.

Sea-rafted pumice with attached serpulid worms. The passengers may reveal something of the journey.

In 1984 pumice from an eruption near Tonga took six months to reach Australia. After another event in the same area in 2001, pumice took one month to reach Fiji and 12 months to reach the coast between Townsville and Sydney. The pumice that washed up in Fiji was clean rock but the pieces that reached Australia carried algae, goose barnacles, serpulid worms, calcareous algae, bryozoans, stony corals and oysters, having had much longer for this collection of organisms to accumulate.

Feathered friends

We would be remiss in talking about shore platforms if we did not have a section on birds and we will again use Long Reef for illustration as it is probably representative of most parts of the coast. We can group the birds into three categories; terrestrial birds that use the shore platforms as part of their habitat, wading shorebirds, and the migrants. Over several years 136 species of birds have been recorded at Long Reef, 68 of them being threatened species.

Locals

On the Long Reef headland an increasing number of terrestrial birds are being seen as the native vegetation is restored. These include the yellow-throated honeyeater *Lichenostomus flavicollis,* white-naped honeyeater *Melithreptus lunatus,* New Holland honeyeaters *Phylidonyris novaehollandiae,* red wattlebird *Anthochaera carunculata*, little wattlebird *A. chrysoptera*, and the variegated fairy-wren *Malurus lamberti*. On the beach you will often meet a willie wagtail *Rhipidura leucophrys,* taking invertebrates from decaying piles of seaweed.

Before World War II the sand dunes were a nesting area for the little penguin (fairy penguin), *Eudyptula minor*. Small, endangered colonies still exist at Manly and Lion Island and occasionally a penguin has been seen on the rocks at Long Reef. It is hoped that dune restoration might attract penguins back but if this is to be successful then we must control domestic dogs, feral cats, and foxes.

Osprey *Pandion haliaetus,* and white-bellied sea-eagles *Haliaeetus leucogaster* prey upon fish near the water surface. The nankeen kestrel *Falco cenchroides*, and peregrine falcon *F. peregrinus* predate small birds along the shore. The Australian raven *Corvus coronoides* is becoming more numerous. It forages on the platform, raids the nests of other birds, and steals golf balls.

Corvus coronoides – not a bird you usually associate with the coast but actually quite common.

One or two white-faced herons *Ardea novaehollandiae* and an occasional eastern reef egret *Ardea sacra* (*Egretta sacra*) can be seen foraging at low tide through most of the year.

Little pied cormorant *Phalacrocorax melanoleucos*, note that this bird has left a deposit on the real estate, all part of the nutrient cycle.

Several hundred crested terns *Sterna bergii* may be huddled together during high tide in winter.

Fifty or so silver gulls *Larus novaehollandiae* will be found feeding or roosting on most days but they move inland during heavy seas.

Sooty oystercatcher *Haematopus fuliginosus*

Most of the high lamp posts in the car park support an Australian pelican *Pelecanus conspicillatus* patiently waiting for local fishermen to bring in their catch whilst casting a supercillious eye on passers-by.

There are usually about 10 great cormorants *Phalacrocorax carbo* drying their wings on the rocks at the eastern end of the headland along with similar numbers of little pied cormorants *P. melanoleucos,* and little black cormorants *P. sulcirostris*.

Long Reef is a feeding site for about 15 to 20 sooty oystercatchers *Haematopus fuliginosus* in late summer and autumn. Unlike most of the other waders these are not migratory but breed locally and can be aggressive if you unwittingly approach their nest that is nothing more than a shallow scrape on the beach. *H. fuliginosus* is a threatened species, but does it catch oysters?

Well, oysters are not so hard to catch; the real difficulty is opening them! But it seems that our sooty oystercatchers are more partial to limpets than oysters and one of us has observed a single bird devouring 88 limpets in a little over one hour. Their attack strategy is either to hammer or chisel the shellfish off the rocks; the preferred technique depends on which one each bird learnt when young. There are members of this genus all over the world and they are important top predators on sandy and rocky shores. Observation of their feeding shows that they take cockles, mussels, pipi, wedge shells, limpets, snails, shrimps, juvenile flounder, and even cunjevoi, but rarely oysters.

Not so local

The double-banded plover *Charadrius bicinctus* has the distinction of being an east west migrant arriving from New Zealand in winter. It develops breeding plumage on Long Reef in August and September and then heads back to Aotearoa for summer.

Visitors from Siberia and Alaska

Many parts of coastal Australia are important feeding or breeding grounds during our summer for birds that have migrated from the northern hemisphere. Most of these visitors are waders that arrive between August and September and begin to return in February. They forage at low tide and gather in roosts at high tide. The red-necked stint *Calidris ruficollis,* is a typical annual migrant from Siberia and north-west Alaska. This is the smallest of the migratory shorebirds with a mass of just 25 gram. In a lifetime of more than 20 years a single bird would fly nearly 500,000 kilometres. It is estimated that some 270,000 red-necked stint can be found around the Australian coast and as many as 160 birds spend our winter on Long Reef. Stints feed on amphipods (beach hoppers) on the rocks and when feeding in sand and mud they use their very sensitive bill tip to 'feel out' prey. As with all 'tactile feeders' the bill is covered by a keratin sheath called the 'rhamphotheca' that has a number of tiny pressure-sensitive pits in it. As the prey species moves away from a probing bill the bird is able to sense the direction of movement and on the next probe, lunch is taken.

The Pacific golden plover *Pluvialis fulva* is a larger wader that breeds in western Alaska. About 9,000 of these spend summer on the Australian coast with 25 to 30 on Long Reef.

The ruddy turnstone *Arenaria interpres* is another visitor from Siberia. Yes, it does turn over stones to feed on invertebrates beneath them, and 20 or so of the 20,000 ruddy turnstones around Australia use Long Reef.

The grey-tailed tattler *Heteroscelus brevipes* and the wandering tattler *H. incanus* (*Tringa* spp.) both breed near stony riverbeds in Siberia. Almost the entire global population of grey-tailed tattlers spend our summer on the Australian coast in mangroves or on rocky headlands.

Short-tailed shearwaters *Puffinus tenuirostris*, and wedge-tailed shearwaters *P. pacificus* are often seen, or dead ones may be washed ashore after a 15,000 kilometre flight from the Aleutian Islands (in the Bering Sea between Alaska and Kamchatka) in September. Unlike the waders these birds breed here in burrows on Bass Strait islands, and parts of the Victorian and Tasmanian coasts where individual colonies may contain up to three million burrows. The young are cared for by both parents until the adults depart for Alaska in mid-April to be followed by the new generation in early May. The flight takes six to eight weeks. Shearwaters have an average lifespan of 15 to 19 years but individual birds are known to have lived for 38 years. That bird would have flown more than 1.1 million kilometres in its lifetime and never received a single frequent flyer point.

One of about 500 dead shearwaters seen along 2km of beach near Batemans Bay in December 2010.

The maritime wanderers

Five species of albatross, most often the black-browed albatross *Diomedea melanophris* have been recorded feeding offshore from Long Reef after flying north to avoid the sub-Antarctic winter.

And tomorrow?

Whether they know it or not our feathered migrants are subject to a number of International treaties, including the Convention on Migratory Species, the Ramsar Convention, and specific agreements between Australia, Japan, China and the Republic of Korea all now tied to the East Asian - Australasian Flyway Partnership. Whilst these agreements are signed off at the highest level of government they must also have practical outcomes in the conservation and enhancement of habitat and the protection of species on the ground. The conservation status of Long Reef makes a small contribution but it is up to all of us to ensure that the chain of quality bird environments all the way to Siberia and back is kept intact.

It is worth noting here that our shore platforms are often the first point of contact in Australia for migratory visitors, and after such long flights they are in desperate need of a feed to replenish energy and increase their body weight. Any disturbance that limits their low tide feeding time can be critical to survival and this includes unleashed dogs. The dogs can't catch the birds but they do love to chase them. Flocks can be chased off, and then have to search for another feeding site while the tide is out. Otherwise they might have an 8 to 12 hour wait for the next low tide. That short period could be the difference between life and death.

Age and longevity

'How long does it live?' This question is asked of us all the time, and most of the time, as with so many other questions relating to the marine world, our answer is, 'We don't know'.

For species that are of commercial significance such as the rock oyster, we do have some answers because life history information is needed for successful husbandry. In commercial fishes we often have a good idea and that helps to regulate the catch. But because much of the data is collected by counting growth rings on fish scales or in sectioned ear bones (otoliths), there is usually some uncertainty about the figures.

Sydney rock oysters *Saccostrea glomerata* live for two to three years before being marketed. The introduced Pacific oyster *Magellana gigas* gets larger in the same time. If left to grow on, both species probably only live four to five years.

Saccostrea glomerata growing happily on the open coast but commercial oyster farms have been destroyed by viruses in many NSW estuaries

Blacklip abalone *Haliotis rubra* reach a shell length of about 25mm at one year. At six years *H. rubra* is about 145mm long. It does grow faster on better sites but its maximum length is in the range 150 to 220 mm and some individuals may live to 15 years.

An indication of how unreliable our estimates may be was gained by one of the authors who happened to be in New Caledonia when the Noumea Aquarium had its six-monthly cleaning of the water supply tank. Large molluscs were found in that tank that in other circumstances would have been thought to be several years old. But in fact they could only have been six months or less in age. The larvae must have been pumped into the tank where conditions turned out to be perfect. It was totally dark excluding any growth of algae, there were no predators, and there was obviously sufficient food. That is of course a special case, but it illustrates a few of the factors that make age estimation so chancy.

Most cephalopods apparently have short, sharp lives. All those of one cohort (the same generation) are born together, live together and often die together in mass mortality events. And the longevity of those species a metre in length, such as the giant cuttlefish *Sepia apama* match those of a few centimetres such as the blue-ringed octopus *Hapalochlaena* spp. in living only months or at most a few years. The cephalopods mate only once and the males all die very soon afterwards. The females live a little longer to look after the eggs and then they die. It has been said that reproduction is the cause of death!

Horny beak of a large squid washed up on the beach.

Limpets live for about two to three years, and sea hares usually only live one year, from larval stage in the offshore plankton to reproductive adulthood.

It seems then that most of the shore platform organisms live for only a short time, but there is anecdotal evidence that hints at a relatively long life (in years) for some. Giant clams *Tridacna gigas* on the coral reefs of northern Australia apparently reach sexual maturity at seven to eight years and are believed to live for more than 30 years. Some anemone are known to live for decades in aquaria and there is some evidence that one species of anemone on the Sydney coast may live for a century.

If an anemone buds off genetically identical young, and lives to do that again, and again, and again – can it be called immortal? A similar question arises when trying to estimate the age of seaweeds. In the case of kelp growth is known to be seasonally variable and occurs in a short part of the plant just above the junction of the stipe ('stem') and the leaf blade. Common kelp *Ecklonia radiata* can grow 70cm in summer and 50cm in winter. But we rarely see individual strands longer than about one metre, so at any one time the visible plant is about 12 to 18 months of growth with the total length being limited by loss of tissue at the tip as it breaks off or is consumed by something. Another way to view this is that the plant might be ancient when the holdfast eventually fails.

And the longest-lived invertebrate species is? Well we are not sure, but measurement of ^{14}C (radioactive carbon) from atmospheric nuclear testing supplemented by growth studies on the east Pacific (USA) red sea-urchin *Strongylocentrotus franciscanus* showed that they could live and continue to breed to 100 and perhaps to 200 years.

The age determinations of *S. franciscanus* depended on the accidental tagging of sea-urchins with radio-active fallout. A few comparable studies have been undertaken in Australia on the common sea-urchin *Heliocidaris erythrogramma* but these used the deliberate introduction of a chemical tag. Large numbers of sea-urchins were captured, their test diameters measured and they were injected with a chemical that became incorporated in the growth lines on their teeth.

The urchins were recaptured one and two years after this treatment. They were measured again and the teeth (Aristotle's lantern) dissected out. The growth line with the chemical tag could be identified under ultra-violet light and that provided a scale for measuring and counting all of the other growth lines in the teeth.

When plotted on graphs the age/growth curves were fairly consistent in different habitats in the same environment but rather different when comparing eastern Tasmania and NSW. The numbers show that most of the sea-urchins lived for 12 to 15 years, quite a few to 25 and 30 years, which was the time when the urchin test stopped growing, and a small number of lucky individuals lived to about 40 years.

As aged scientists ourselves we wonder if the sea-urchins think that is lucky?

A myriad ways to die

Death on a shore platform comes in many different ways.

Predation

To become lunch for something is probably the norm for most organisms. All of the usual predator/prey relationships could be discussed but as we have said quite a bit about what eats what elsewhere we will keep this paragraph short. Occasionally a predator or grazer population will suddenly get out of control, and then it can have a devastating effect on its prey. The crown-of-thorns sea-star on the Great Barrier Reef and elsewhere is an example. In temperate waters we can cite ‘blooms’ of jellyfish, or the rapid expansion of ‘barren grounds’ created by large numbers of sea-urchins. The latter are prominent in shallow waters of eastern Tasmania where large patches of the sea floor are denuded of algae and thus all other life forms by high densities of invasive long-spine sea-urchins *Centrostephanus rodgersii*. This problem seems to have increased since the 1950s and may also be an early indicator of climate change.

Disease and parasites

In a sense disease organisms and parasites are another form of predation and under normal conditions they will not kill all of the host species. It’s tough on those that die and there are circumstances when the kill rate is so high that population structure and size can be significantly affected. These events most often involve the arrival of a new disease or parasitic organism to which the hosts have no effective defence, and there is some evidence to suggest that such incidents have been increasing in the marine environment in recent decades.

Storms

Extreme coastal storms, such as the events of May and June 1974 on the east coast can seem to sweep a shore platform clean of organisms. However, enough individuals do survive to repopulate and in time new intertidal communities establish.

In many places the 1974 storms completely removed the Neptune’s necklace *Hormosira banksii*. Where the holdfasts were less damaged it quickly returned but where holdfasts had been destroyed new communities arose that included dense patches of the honeycomb barnacle *Chamaesipho tasmanica*. Experimental work to determine why this occurred found that it was actually the loss of mulberry snails *Morula marginalba* that had previously lived beneath the seaweed that allowed the barnacles to take over. It was suggested that *Hormosira banksii* was a keystone species in providing canopy protection for so many other organisms that the loss of algae caused a cascade of unexpected events in the recovery phase.

Desiccation

Drying out is one of the more important mechanisms culling organisms at the upper tide levels and contributing to zonation. Occasionally a more extreme drying event occurs when there is a combination of daytime low tides, low swell and high air temperature. This is called a 'conjunction of rare events' and probably happens with the same frequency as big storms, say once in 10 years with the effects being equally important.

Burnt *Hormosira banksii* exposed to air temperatures above 35°C at low tide.

Pollution

In mentioning pollution of coastal waters we usually think of oil spills. There are several appalling examples on record, such as the *Torrey Canyon* stranding at the entrance to the English Channel in 1967, and the biggest maritime environmental catastrophe of the *Exxon Valdez* spill in Alaska in 1989, where attempts were made to clean up rocky shores with detergent and steam jets. Environmental scientists learnt a huge amount from these accidents and reports about the recovery of coastal organisms are still being made. One estimate is that the Alaskan shoreline would not return to something approaching its original biodiversity for at least 30 years after the spill.

The biggest ever (?) oil pollution event occurred in 2010 when the Deepwater Horizon well in the Gulf of Mexico blew out and leaked nearly 10 m litres of oil every day for three months. Cleanup costs exceeded US$65 billion.

Oil spills are not unknown along our coast. In March 2009 the *MV Pacific Adventurer* spilt 270,000 litres of fuel oil north of Moreton Bay. It cost about $60m to clean up, and whilst the media reported its effect on beaches, they never mentioned shore platforms.

Even more insidious and occurring with every rainfall are 'minor' pollution events coming from stormwater drains that discharge on the beach or the shore platforms. These flows carry organic materials, oil films, rubber particles, and high loads of heavy metals, especially zinc. Zinc is a potent pollutant and in these cases it comes from – wait for it, - car tyres. Coastal contamination by stormwater runoff in urban areas rarely gets the management attention it requires.

Holes in shells

Any handful of dead shells from the shoreline will always include some with holes in them. Often those holes are randomly scattered, and made mechanically when a wave bashes a shell against a hard surface. Most of the larger limpet shells found on the strand line have very corroded peaks, and if you hold one up to the light you can see how thin they have become in comparison to the edge of the shell. Search a little harder and you will soon find ‘limpet rings’ formed when the thin portion has broken away from the main shell. We do not know what causes this but it appears to be an etching process rather than abrasion by moving sand, and we suspect the glues involved in attaching algae and other organisms. Does this simple process control limpet size and age?

Shellfish form their shells by adding new organic matter and mineral at the edge of the mantle around the opening. First a polymer of protein and chitin (conchiolin) is deposited, then a prismatic layer of calcite followed by a pearly layer (nacre) of aragonite. Calcite and aragonite are both calcium carbonate minerals with well-developed cleavage and are easy to break. They have different crystal structure and as each mineral layer is orientated differently, they reinforce one another and a crack in one layer is unlikely to propagate through all three.

A limpet ring and the eroded or corroded shell from which they are derived.

Bored holes look very professional, with neat, chamfered sides, and they often occupy the same spot on many individual shells. The animals once resident in these shells have all fallen victim to a predator that attacks them by drilling the hole.

One family of gastropods seems dedicated to drilling holes. Whelks are the experts, good examples being the cartrut shell *Dicathais orbita*, and the mulberry snail or black oyster borer *Morula marginalba*, a species heartily disliked by oyster farmers for polishing off many an oyster. The name *Morula* is derived from the Latin for mulberry *Morus,* and this little fellow is about the same size and does look remarkably like that fruit. It is widely distributed even into the muddy estuaries, provided that there is a bit of rock around. Despite the other common name it is not solely a predator of oysters and is quite partial to barnacles, limpets, and tubeworms.

On shore platforms *M. marginalba* shelter in clustered groups in small depressions during low tide in warm weather. They are more active in cooler weather and disperse across the platform seeking food. Commonly there is a halo of empty space around any one cluster in which no prey species are found, so this little carnivore has a very real effect on patchiness on the shore platform.

Predatory snails like these can be distinguished from grazing snails by the presence of a siphonal canal at the tip of the shell. The siphon protrudes through this groove and senses prey as the snail crawls across the substrate. On finding lunch the whelk sits on top of the victim and softens the shell by secreting an enzyme from a specialised gland in its foot. This gland has been labelled the 'accessory-boring organ'. There is nothing like giving something a fancy name; it covers our ignorance! When the shell surface is softened the drilling process starts using mechanical abrasion by the radula. These processes are repeated sequentially and observations show that it can take up to four days for *M. marginalba* to drill through a barnacle. This gives new meaning to 'having a long lunch'.

After drilling the hole the snail inserts its proboscis, secretes enzymes into the flesh of the organism and sucks out the broken down organic matter. *M. marginalba* is in turn preyed upon by birds, crabs, octopus and some fish.

Holes drilled in limpets by *Dicathais orbita* are not in the thinnest part of the shell, but located over the gonad and digestive gland of the prey species. These holes also happen to be in a position where the whelk fits neatly onto the limpet shell. This raises a question: do the whelks do this to gain access to energy rich foods, or is it that holding on to the prey whilst drilling is more important?

Bivalves, *Glycymeris grayana*, drilled by a naticid gastropod.

Not all boring species target richer food parts. Some locate their drill site on thinner areas of shell, and when a whelk *Lepsiella baileyana* (*Haustrum baileyana, Thais baileyana*) feeds on abalone it drills through thick shell to reach muscle rather than gut. If you've ever eaten abalone you'll appreciate this selection. These guys take up to seven days to drill a hole and can then feed on the abalone for another seven days, a very long lunch indeed.

The family Naticidae does not have a siphonal canal. It is a cosmopolitan family of mainly intertidal or shallow water molluscan carnivores that are predators on other molluscs, mainly bivalves. Masters of their art, they too seem to know just where to drill; their food is at the other end of that process, and a badly positioned drill hole will be wasted effort. With bivalves, they target the little knob near the hinge between the shells (the umbo) as there they are assured of piercing into the visceral organs. Once the hole is complete they kill or incapacitate their prey, and then proceed to suck out the organs whilst carefully enveloping their capture with their water engorged foot.

What about a shell-drilling octopus? Unbeknown to most people octopus are expert borers. When a mollusc proves too strong for an octopus to pull open, it selects the right spot to drill, and with its own version of an 'accessory-boring organ'; it begins the slow drilling process. The hole made is not neat like that of a naticid. Close inspection of an octopus hole reveals the crude shape of a figure eight, with rough sides. Octopus seem to love cowries (family Cypraeidae) and often every second dead shell will show a very distinctive hole, always on the same spot over the visceral mass. A cowry shell has a very hard-to-get-at animal tucked away behind those two rows of 'teeth'. The octopus drills, injects saliva and externally digests the animal. Occasionally, a dead cowry shell will have a number of randomly placed holes. This most likely represents the activity of a young octopus, still learning.

Even some sponges are capable of boring holes. These are the minute holes you may see on a shell surface. At times these sponges attack the exposed shell, perforating it to such an extent that the animal has to rebuild the shell. Sponge attack is a major cause of the degradation of the shell with many of the smaller fragments ending up in beach sand.

The ultimate fate of any shell is to be ground down to beach sand and then

Cowry shell showing evidence of an octopus having snacked.

dissolved. It should be no surprise that the beaches around large shore platforms have the highest proportion of shell sand (calcium carbonate) simply because the shore platform is so productive. Some of the shell sand will dissolve in rainwater. Calcium carbonate can then cement sand around plant roots to form rhizomorphs (root casts), or cement whole beds of sand to make dune rock (aeolianite), or more rarely, beach rock. All of these forms of cemented sand may be seen near Long Reef, Avalon, and at Palm Beach, and in every case there is a wide and productive shore platform nearby.

In this text we have said very little about beaches but as sand beaches cover 60% of the New South Wales coast it would be remiss to leave them out entirely. Beaches are also habitat for marine organisms although we do not often think of them in those terms. But we can compare them to shore platforms in that their living space can be divided into zones in much the same way as a shore platform can and these zones reflect the distribution of wave energy and saturation.

Beaches are often backed by dunes that carry terrestrial communities and in which land animals and birds are normally found. A little lower in sand that is often moved during big storms ghost crab burrows can be found amongst the spinifex.

In the drift line where seaweed and other floating debris is concentrated by waves in the run-up zone above high tide, decomposing seaweeds harbour scavenging isopods, amphipods and kelp fly, all of which provide food for sea birds.

The swash zone is saturated by salt water and true marine organisms like pipi, beach worms, and crabs live there. No light penetrates the sand below a few millimetres depth, oxygen, temperature and organic matter all decrease with depth and there is frequent disturbance of the whole mass. However, the space between sand grains and the saturated mass of sand supports copepods, flatworms and predators like nematodes and polychaete worms. In the thin surface where there is enough light diatoms and dinoflagellates survive.

Below the swash zone the environment is submarine and the higher inhabitants of the surf zone are jellyfish, fish, porpoises along with frequent visitations by humans on surf boards.

We understand the physical dynamics of beaches fairly well because we have been attempting manage them for decades with our main focus on limiting sand loss and property damage by erosion. We have undertaken far fewer studies of beach ecology and many of our beach management actions, such as sand grooming, can have negative effects on the sand dwellers.

Sponge boring on the edge of *Haliotis rubra*. Note the scale.

Shell sand and quartz sand from a Sydney beach. The largest gastropods here are less than 3mm long.

The strandline

No description of the shoreline would be complete without some reference to the strandline and the treasures to be found there. No, we don't mean all the left foot thongs, cigarette butts, and sticky balls of fat from sewage effluent as we are sure you are familiar with these, but we would like to tell you about some of the more unusual things. A number have already been mentioned like pumice, shells of the rams-horn squid *Spirula spirula*, goose barnacles *Lepas pectinata,* and *L. australis,* cuttlebones, egg cases, and bluebottles *Physalia* spp., so lets check some others.

Along the ocean beach after big storms you are likely to find huge piles of sea wrack, a word derived from the Dutch and German languages of the middle ages that originally meant 'a wreck'. And of course we do have plenty of real shipwrecks. More than 2,000 are recorded along the NSW coast and you will sometimes find traces of them amongst the rocks on the shore platform. One example is a rusted tangle of anchor chain on Long Reef that probably marks the point where the iron paddle-steamer *Euroka* ended up in 1913. Built in Balmain, only 16 years old and loaded with coal from Lake Macquarie, whilst heading for Sydney she began leaking and was abandoned. She then drifted onto the rocks, probably close to the point where the *SS Duckenfield* had come to grief 24 years earlier with the loss of one life. All coastal wrecks more than 75 years old are protected under the NSW Heritage Act 1977 and the Commonwealth Historic Shipwrecks Act 1976.

Every now and then very large quantities of seaweed, most often kelp *Ecklonia radiata* and Neptune's necklace *Hormosira banksii,* come ashore during a storm. Is this stuff just ripped up by big waves or is something else going on? In the 1990s dieback in that same species of kelp with more than 90% mortality was observed at several locations in North Auckland, New Zealand. Laboratory studies of dying kelp identified a virus, one of only a few recorded cases of viruses in seaweeds. As usual this discovery raised more questions. Was the virus responsible for the dieback? Is viral dieback associated with other big wrack accumulation events? Is this a natural process or has there been some change in the kelp beds that favoured the virus? Is the same virus, or others, an agent in the mortality of other seaweeds? If there are few viruses in seaweeds relative to their numbers in vascular plants why is that? Do seaweeds have some sort of viral protection mechanism that may be useful in agriculture?

You guessed it, we have no answers but they are all great questions waiting for careful observers and researchers with the right skills.

A big stranding of common kelp at Long Reef in April 2009.

In amongst the wrack are numerous other treasures both dead and alive. Miscellaneous plant parts are common, especially on more tropical coasts where many of the shore plants use the sea to disperse their woody fruits. You may see the germinated fruit of the local grey mangrove *Avicennia marina.* Or plant parts that have drifted from Queensland such as woody seed cases looking like smooth brown golf balls of mastwood or ball-nut *Calophyllum inophyllum,* broken pineapple-like fruit of the sub-tropical screw palm *Pandanus tectorius*, and perhaps the big boxy fruit of a tropical mangrove, the fish-killer tree *Barringtonia asiatica*.

Fruit of *Avicennia marina.*

Flotsam of animal origin is well represented. Fragmented sea-urchin skeletons (tests) are not uncommon, and you may find an almost complete one though they do not usually retain spines or the jaws known as Aristotle's lantern. Dead fish and migratory sea birds may be present, along with the empty

Seed of *Calophyllum inophyllum*, a little smaller than a golf ball.

Fruit of *Barringtonia asiatica* loaded with bryozoans and stalked barnacles.

carapace of crabs that are actually cast-off moults, sponges, numerous shells, and of course any amount of junk.

Beautiful structure of an urchin test.

As the organic wrack decomposes it becomes a source of nutrients to the beach, dunes and shallow water, and is habitat and food resource to numerous decomposer organisms. The most visible of these are the sand-hoppers or kelp amphipods *Notorchestia quadrimana,* easily found when you turn over a bunch of wrack as they leap everywhere. Amphipods are related to lobsters, crabs and barnacles. Their role in life is to break down all the organic matter on the shore, and there are so many of them that they provide a diet staple for shore birds.

A three-barred porcupine fish probably *Dicotylichthys punctulatus* stranded on the beach. Scale in cm.

Adaptation to a particular niche is sometimes played out so well that some organisms have become totally reliant on a certain food resource. Australian kelp flies (Coleopidae) are a good example. Globally there are only 28 species in this family of which 13 occur on the Australian coast. All rely on stranded marine algae, mainly kelp, or seagrasses, on which the larvae feed. Consequently, they play an important role in consuming this food, thus ridding the beach of a sometimes unpleasant and bulky mass of organic matter. At times the adult, black flies mass in such high numbers they appear to be a health threat, but although they cling to passers-by they quickly abandon them as the people move on. Kelp flies are only interested in kelp.

Nutrient flow from the strand line is one of the ecological links between the sea and the land that you can sometimes see in action. Observations by a student at Long Reef have shown an amazing seasonal association with a very common, small black ant often seen foraging along the tide line. Ant activity increases over the summer months, as it appears that the main target is bluebottles with ants swarming over beached specimens. This observation has been confirmed over two years and is a perfect example of nutrient recycling from the sea to the land by an insignificant terrestrial organism that converts this large, non-selective marine predator into just another food source.

For thousands of years the flotsam and jetsam derived from human activities was composed of natural products like wood and fibre. As it broke down this behaved much like any other organic material and although it may have taken longer to decay it eventually did. In the 19th Century or perhaps a little earlier glass products began to turn up and probably peaked in abundance in the first half of the 20th Century when bottles, fishing floats, and light globes became very common.

This is no joke, it really was washed in as flotsam. What story could it tell?

In the last 50 years debris derived from people has become dominated by plastics, and the volume has hugely increased. Plastics of every imaginable colour, composition and purpose can be found, from nurdles to ghost nets, rope ends, shopping bags, drink bottles, packaging and even abandoned scientific experiments. Approximately 60 to 70% of the rubbish floating in the ocean is plastic. Some is dumped at sea, some on the beaches by careless people but a great deal of it reaches the coast through urban stormwater drains and ends up drifting long distances, to be dispersed on coasts around the world. In the process it carries large populations of bacteria, bryozoa, and barnacles. Floating plastics trap and choke fish, turtles, sea birds and dolphins and the problem seems as if it will last forever.

Common seadragon *Phyllopteryx taeniolatus* also washed in as flotsam. A moderate number were recorded at the same time along several Sydney beachs.

Do we really have to make fishing lures, hooks, lines and sinkers from materials that are, indestructible and deadly long after the fishermen have lost them?

Isn't it time we took stock of our bad habits?

Underwater communication

In the terrestrial world vertebrates find mates and food by engaging the five senses; sight, hearing, touch, taste, and smell. In the marine world the relative importance of these differ as sight is limited by water depth and turbidity, touch is of little use if you can't move around, taste and smell more or less combine, and hearing is enhanced by the better transmission of low frequency sound through water.

Whales and porpoises talk, fish serenade one another with an amazing range of sounds, snapping shrimp noise levels are loud enough to confuse the navy and even jellyfish respond to sound.

Different marine environments have distinctive sound profiles, a coral reef sounds different from a seagrass meadow and motile or migratory species and perhaps even larvae can differentiate these and move in the direction of their preferred habitat.

Too much underwater sound generated by human activities becomes noise pollution and this can cause temporary or permanent hearing loss in marine creatures, increase stress and disrupt navigation, feeding, and breeding. Excess noise is implicated in mass whale strandings. Fish populations can be different near busy ports and sea lanes and sound can even be fatal. A single airgun blast during seismic survey has been seen to kill all the larvae in the water for a distance of 1,200m around the blast point.

A healthy coral reef is the loudest ecosystem in the sea and such reefs have been shown to attract more fish. Quiet reefs signal lower species diversity and perhaps degradation.

Ocean waters teem with chemical signals called pheromones, that affect individual species behaviour and that can determine; settling, foraging, fight or flight responses, as well as mate selection. On the larger canvas pheromones are very important in affecting community structure and would you believe it, energy and nutrient flow and climate change. For example, when certain phytoplankton chemically sense their neighbours being attacked by ciliates that feed on small foods, the phytoplankton change behaviour and grow as colonies. When those colonies are attacked by copepods that feed on larger food, they revert to single cells that are too small to be of interest to copepods. The chemical signals in this case are the 'smell' of death.

What chemicals are involved? Probably not a huge number. Sugars, proteins, amino acids, and metabolic waste products signal food and even humans are familiar with those. There are also very particular chemicals that can carry distant messages such as dimethyl sulfide (DMS). DMS is released when zooplankton attack blooms of phytoplankton, and it is found in the excreta of many predators. It is described as a repulsive odour reminiscent of boiled cabbage. Tube-nosed seabirds such as petrels and shearwaters can sense DMS concentrations perhaps as far as 100 km away. High atmospheric DMS points to high marine productivity and the birds fly in. DMS is also dispersed in the water body and attracts predatory fish.

Some pheromones give chemical warnings to potential predators about taste or toxicity. Decorator crabs take advantage of this and select nasty-tasting seaweeds, anemones, sponges, or bacteria to transplant to their carapace and thus gain protection. Some of the chemicals involved have been identified as has the glycoprotein that cues barnacle larvae to settle within reach of other barnacles.

A study of pheromones released by sea hares (*Aplysia sp.)* during egg laying identified four proteins that are resistant to rapid breakdown. The biologists involved named them; attractin, temptin, enticin and seductin. Who said biologists were humourless?

Pheromones play multiple roles in sex, mating and fertilization from bringing individuals together, to stimulating or deterring copulation when in contact, and even promoting or preventing fertilization when egg and sperm are close. Gastropods find mates by following chemical signals in mucus trails and it has been shown that pheromones are so powerful in the mating behaviour of some crabs that they can be tricked into copulating with golf balls, well trying to anyway, if the balls have been treated with female crab urine that carries the signal.

The study of pheromones is relatively new but even so this chapter could easily stretch out to a book full of examples but rather than do that let us go back to DMS. We know little about chemistry but we have heard of this molecule elsewhere.

In the 1970s James Lovelock a chemist, and Lynn Margulis a biologist, presented the Gaia hypothesis which suggested that living organisms interact with their inorganic surroundings to form a self-regulating, complex system that can maintain and perpetuate the conditions needed for life. Their model treated the planet itself as a self-regulating super-organism. The hypothesis generated a lot of discussion and criticism, but it also led to much research seeking to illustrate the mechanisms. One program was investigation of the role of DMS released by oceanic plankton and its influence on climate. DMS breaks down in the atmosphere to sulpher dioxide which is important in cloud formation. At that time the only known sources of sulpher dioxide were volcanic gases and industrial pollutants but the numbers did not add up until Lovelock suggested that DMS was a missing source. After adding the numbers, it could be seen that biological regulation of the climate was possible through phytoplankton growth and liberation of DMS and that the links between solar energy input to the oceans, production of DMS, expansion of cloud cover, and increased solar energy reflected back into space (albedo), was a closed feedback loop that was sufficiently powerful to stabilise global climate.

Drilling down a little shows the story going further. The geographic distribution of DMS attracts sea birds to richer food resources and they in turn return to the land to roost where their faeces transfer and concentrate phosphorous into the soil that can support particular vegetation. A local example is forests of *Araucaria heterophylla* on Norfolk Island where phosphate levels were a major ecosystem control when huge numbers of Providence petrels *(Pterodroma solandri)* roosted amongst them. That bird was eaten by the colonists and driven to extinction by 1800. Today phosphorous levels have declined and many trees are affected by root-rotting fungus and are dying. That nutrient cycle was broken 200 years ago but is only now evident.

In this story we have connected the production of one smelly gas to dense coastal forests, abundant phytoplankton and oxygen generation, then back to deep cold waters with high nutrient loads brought to the surface by eddies in an ocean gyre.

What a complex world we live in and we have just opened a tiny window into it.

Shellfish reefs

Of all commercial shellfish, oysters would have to be the most popular. Perhaps because they are said to be an aphrodisiac although we suspect the evidence supporting this is pretty thin! Oysters are widespread on rocks and the trunks of mangroves in tidal waters of the east coast and have been commercially harvested through a chequered history of success and failure worthy of a Shakespearean play.

When James Cook landed in Botany Bay, he found the Gweagal people were collecting oysters as a food staple. Archaeological evidence from Kurnell revealed that they had been doing that for 2,000 years and no doubt much longer, but the evidence for earlier use has been lost to sea level rise.

The Australian flat oyster (*Ostrea angasi*) occurs as a bottom dweller and shellfish reef former in the shallow water of the protected bays and estuaries. Its round and rather flat shell closely resemble the European oyster (*Ostrea edulis*) and grows to 150mm. Europeans immediately recognised it as a familiar food, consumed in quantity at the Globe Theatre in Shakespeare's time.

L to R: A waterworn example of *Ostrea angasi, Magallana gigas* and *Saccostrea glomerata*. In all the species one valve is much flatter than the other.

The Sydney rock oyster *Saccostrea glomerata (S. commercialis* and *Ostrea commercialis),* attaches to hard surfaces along the east coast to a depth of 8-10m.

In pre-European times both flat and rock oysters formed extensive shellfish reefs with oysters growing on oysters for generations and in the process contributing numerous environmental services such as water filtration, shore protection, nutrient cycling and fish habitat. Rock oyster reefs typically carried >500 live shells on every square metre. In South Australia alone the total length of reef before settlement is estimated at 1,500km and the Australian total far exceeded the dimensions of the Great Barrier Reef. Today there is less than 1% functionally intact.

Flat oysters were extensively collected by early Europeans both to eat and to convert the shell to lime for mortar. Collection involved dredging the entire reef resulting in destruction so great that by the 1860s stocks were scarce and the government closed harvesting in the Georges River in the hope that the reefs would recover. By 1890 flat oysters were deemed to be extinct in NSW End of Act 1.

Former saltmarsh in Gwawley Bay in the Georges River estuary showing traces of Holt's 1872 attempt at oyster cultivation. The canal lines visible at the time of the photo (1943) were then colonised by mangroves (NSW Spatial Services)

Harvest closure and the degradation of oyster reefs led to passing of the NSW Oyster Beds Act (1868) that prohibited collection and burning of live oysters for lime manufacture. Oyster cultivation seemed like an answer and the politician Thomas Holt (1811-1888) constructed an extensive network of ponds and ditches (*claires*) on saltmarsh flats in the Georges River where nearly half a million sandstone blocks were laid to provide a substrate for rock oyster spats (attached larvae) in 1872. This attempt failed because the shallow waters were too warm and had too much suspended sediment. Eventually the more successful stick and tray system evolved that kept the oyster above the mud and out of the water for about 70% of the time. Exposure during low tide helps to control mudworm *Polydora websteri* infestation. The mudworm first became a problem after oyster spat were imported from New Zealand between 1889 and 1898 when local supplies were depleted. Was there a connection between these events? Perhaps, but proof is lacking.

Today that entire bay has been entirely changed by landfill and urban canal development. End of Act II.

The same area in 2020 now called Sylvania Waters. No saltmarsh, few mangroves and oysters long forgotten (NSW Six Maps).

The life cycle of *Saccostrea glomerata* is pretty simple. They are filter feeding bivalves that mature at three years in the wild. Food consists of virtually any organic matter in the water. High loads of sewage effluent enhance growth and oysters can accumulate heavy metals from polluted waters. Both of these factors present health risks to humans.

Each oyster begins adult life as a male and releases hundreds of millions of sperm in the late summer. Older oysters are female and release 20 to 40 million eggs. Fertilization takes place in the water and the larvae swim for about a month before being attracted to other oysters by sound and chemicals then settling as spat on hard clean surfaces. The survival rate is 0.1% but larvae are abundant.

Annual NSW production of cultivated Sydney rock oysters peaked at 15 million dozen in the 1970s, dropped to 8 million in 2003, then 6 million in 2007 and is still in decline despite research and development of improved strains. Today there is no production from the Georges River and Botany Bay and very little from the Hawkesbury River, areas that were once the core of the industry. End of Act III - what happened?

In the Georges River through the 1960s the local community resented the total area of water surface that was occupied by oyster leases because they reduced recreation opportunities. In the 1970s consumers became ill after eating oysters that had been thriving on sewage effluent. Government requirements to put all oysters through a cleaning process before marketing addressed that issue – at a cost to growers.

In 1968 a mystery disease began killing oysters in Moreton Bay. It was labelled QX - Queensland unknown. It reached the Georges River in 1994 and wiped out all stock. The Hawkesbury River seemed to be free of the disease but it turned up in 2004. The organism was identified as a protozoan parasite, *Marteilia sydneyi*, which reaches maturity in the oyster and apparently has an intermediate growth stage in a local polychaete worm, *Aglaophamus australiensis (Nephtys australiensis)*. The Sydney industry was virtually wiped out by 2005 and growers adjusted by using disease resistant stock, expanding operations in Port Stephens where the water was cleaner, and farming Pacific Oysters *Magallana gigas (Crassostrea gigas)* which was previously regarded as an illegal immigrant and declared a 'noxious fish' in all NSW waters other than Port Stephens. The law can be a very blunt management tool!

Between 2010 and 2013 Pacific Oyster Mortality Syndrome (POMS) turned up and killed all oysters in infected areas in days. This is an oyster herpes virus that may also be hosted by another polychaete worm, but as it acts in conjunction with a number of bacteria working out the real cause(s) of death is a challenge. In 2021 QX was detected in Port Stephens. Biosecurity regulations were invoked and there are now tight restrictions on the movement of oysters and equipment.

Is this the end of the tragedy, or just Act IV?

There has to be some good news in this story and it can be seen in current attempts to re-establish oyster reefs. The first trials began in Port Phillip Bay in 2015 and 46 projects are now underway from Queensland to West Australia and others proposed in Tasmania. The total area involved is tiny, and different techniques are employed but results are impressive on these community driven demonstration sites and the idea is going international through the Australian Shellfish Reef Restoration Network.

Next time you ingest a humble oyster, or stub a toe on those sharp shells consider the chequered history this species has had in Australian waters with our clumsy efforts to utilise it. Effective management and quarantine have often been secondary issues to immediate commercial return and our ecological understanding of the species is still limited.

Perhaps we should let William Shakespeare and King Lear (Act 1, Scene 5) remind us of how little we know:

Fool	Canst tell how an oyster makes his shell?
King Lear	No
Fool	Nor I either, but I can tell why a snail has a house.
King Lear	Why?
Fool	Why, to put his head in; not to give it away to his daughters, and leave his horns without a case.

On learning more

By now you will have begun to realise how little we know of the watery world along our shore, and you will have seen that every time we answer a simple question several more emerge. This is the way of natural history, and if we have inspired you in any way, you are perhaps now grumbling about our lack of references and itching to learn more.

Family exploring between the tides.

This little book is crammed full of information, but where did it all come from? Yes it is stored in a score of textbooks, but the professional authors of these works did not discover it all, they compiled and interpreted the observations of hundreds of others who often had no academic training and had simply seen something interesting to which they had applied their own explanation. Our knowledge of the natural world has its foundations in the observations of ordinary people.

Something sparked their interest. It could be you, as a fisherman, using beach worms as bait, and wondering why those fellows who caught the worms referred to them as bronzies, red heads, stumpies, white heads, or green heads, and yet the experts said they were all one species. You thought, 'That couldn't be right', and after enough pestering one expert decided to take another look, and finally showed that they were several different species.

Biologists getting down to it in the search for an elusive shore platform organism. Yes, they did put the rocks back carefully.

That happens often, because 'you and me' are roaming shore platforms and beaches all the time, and we notice differences that the 'experts' have missed.

One of the authors of this book, has been puzzled for years about a seemingly random but very noticeable act of what might be called biological vandalism, with literally hundreds of limpets flipped on their backs, at low tide. Of course limpets cannot turn themselves over, and will die in ten minutes when exposed to the sun while upside down. With continued observation and much help from a bird expert, it looked as if one bird species, the migratory ruddy turnstone *Arenaria interpres*, was the culprit. The theory is that these birds like eating limpets, but for some reason the *Siphonaria* limpet that is so common on the rock flats is unpalatable. So any limpet seen is turned over, but *Siphonaria* are abandoned, while other species are quickly devoured.

This story immediately raises more questions. If this has been happening for years, decades, centuries, or more, you would think that there would be no *Siphonaria* left. But at certain times of the year you will find literally millions of very small *Siphonaria* appearing everywhere. Has this predator/prey relationship between birds and *Siphonaria* been going on for so long that evolution has shaped a solution to this mass mortality? Perhaps it is only the new, young birds that have not yet learnt to discriminate between edible and inedible limpets? Or what?

For every story we solve, ten questions immediately pop up. This is what is so exciting about studying marine biology. There are so many things that are waiting to be discovered, and you can be right there on the spot to solve the puzzles.

On the way through this book we have included photos of supposedly rare animals. We class them as rare, but it could be that in their favourite habitat they are quite common, but that habitat has eluded us. Or it could just be that they have a short season as visible adults and come together to breed and then disperse. Often the nature of their young or eggs is a mystery. Your observations might solve some of these questions. Take notes, and photos. Read books. Find an expert at a museum, university or school. Join a relevant club or society. You just might be lucky enough to make that critical discovery and if you do then that is the moment that you should contact the experts and drag them away from their computer screens. They too will be excited, and by the way, they don't bite ☺

Tomorrow's biologist in training?

By way of example, we can reveal a few recent discoveries.

Whilst looking in tidal pools at Turimetta a local observer and his daughter found blue anemones. They checked the guide books seeking a name, but with no result and they brought the problem

But will she have good shore platforms to work on? Seascapes like this at Narrabeen Head are likely to change substantially as a result of sea level rise in her lifetime. Photo by Peter Donkers

to us. We had to plead ignorance. The only example of blue anemones we could find were species in the United Kingdom. We still don't have a name for this one but we did learn that the common species *Anthothoe albocincta* exhibits a considerable range of colours and that *Oulactis australiensis* can have shades that are almost blue. We know that groups of anemones are often genetically identical (clone groups) and as this particular find was close to a sewage outlet that might provide an explanation.

Sometime later the same observer reported large trace fossil tracks in rocks exposed when beach sand had been eroded by storms. This was exciting as nothing so large had been reported in the geological literature even though there is an entire PhD on trace fossils in these rocks. We came up with a possible explanation, made a short video about it, then called in palaeontologists from the University and Museum. They argued for a different explanation and we made another video (Pittwater Pathways). This is worth viewing as it illustrates

A new species or a rare clone? Recently discovered blue anemones at Turimetta. Photo by Paul Cronk.

Recently discovered large fossil tracks in the Bald Hill Claystone made by – what? Photo by John Illingsworth.

how scientists seek answers – but further thought points to both explanations being wrong! The story is a work in progress as the tracks have gone back under the beach and will not be seen again until the next erosion event.

There are always new questions, anyone can raise them, and answers are often slow to be revealed. We have puzzled over some questions for years and are still waiting for answers. For example, why do black nerites *Nerita atramentosa* locate in full sun at the highest tide level even in the middle of summer? But perhaps the better question is not why but, how do they survive the thermal load?

We will finish with three other stories that may have commercial outcomes.

Tidal swimming pools on the coast have a constant problem with algal growth making floors and walls slippery and dangerous. Until recently the easy way to keep them clean was to give a regular wash with chlorine. But that was flushed into the adjacent open sea and has been an environmental disaster for everything living even some distance from the pool. By studying what eats that algae (periwinkles or limpets); by monitoring movements of those suspect periwinkles or limpets; by observing their habits such as clustering between tides (why?), or sheltering after feeding (where?), we have potential management tools. If those suspect gastropods need shelter, then provide it by designing ledges or making the pool walls rough to provide suitable micro-topography. The problem may be simply solved and what is more the answer could be self -maintaining.

Boat hulls are constantly 'fouled' by attached marine organisms, as the numbers accumulate and the hull community develops the boats are slowed in the water and boat owners seek a remedy. There are two obvious answers, prevent the fouling in the first instance or regularly clean the hull. Paints containing anti-fouling compounds have long been used and some of the most successful chemicals have contained compounds of tributyltin (TBT). As early as the 1960s this was found to be leaching into the marine system and having adverse effects on local fauna. It is also toxic to mammals including humans but its use was not banned in Australia until 1991 (for boats under 25m length) and 2008 (for all boats). Obviously we need a safer approach to antifouling and the answer may lie on the shore platform. Why is it that most sponges do not support other organisms on their surface? The answer is that they are able to shed their surface regularly and thus dislodge anything that has settled there. So should we be thinking about self-peeling paints?

The slip, slap, slop, routine of sun protection is normal for most of us in summer. Modern sunscreens work remarkably well but questions have arisen about the safety of one of their ingredient, nano-sized particles of zinc. Maybe it is safe, maybe it isn't, the jury is out. But what protects inter-tidal organisms and particularly their eggs from sunburn and could we develop new products from them? You have to ask the question before any answers are evident and that is exactly the question that a Wollongong post-graduate student asked herself whilst working on gastropod breeding cycles on the shore platform. She found that the egg masses of a number of local gastropods including the denticulate siphon shell *Siphonaria denticulata* (a false limpet) contain an agent that protects them from harmful ultra-violet radiation. Her answer may well lead to a new natural product for the slip, slap, slop routine in a year or two.

Three simple examples where observation of the ordinary or the esoteric, have led or may lead to entirely new ways of dealing with practical problems. How many more answers are out there? We will never know unless we look and unless we ask the right questions.

Rethinking shorelines

Geologically we live in the Anthropocene, a new geological time unit that is soon to be formally defined by the International Commission on Stratigraphy (in their wisdom tthey decided otherwise in early 2024). It is the time when humans must be recognised as a global power exerting major influences on the functioning of Earth Systems. Nowhere is this more apparent than at the coast where the rate of change is greater than the Earth has seen in the last 18,000 years.

On our coast the EAC is stretching further south, the water is warmer for longer, tropical species are appearing more frequently and surviving, severe storms seem to be more frequent, sea level is rising and may easily reach or exceed 1m above the present level by the end of this century, and the coast is responding. We are increasingly concerned about erosion and the loss of beaches. Severe storms are commonly described as 'the worst in living memory'. That claim is actually hard to justify as we do not have good meteorological data to support it, but what is true is that the next storm almost always does more damage than the last one of similar magnitude. Why? Because we build more and more expensive property in risky places.

We apply engineering solutions to beach losses with the most ostentatious local attempt being a $25m concrete sea wall at Collaroy but will it protect the property behind it when sea level is up 1m and we get a 1:500 storm? Can we afford to apply such partial answers to beaches everywhere?

Collaroy-Narrabeen Beach immediately after the 2016 storm. Photo by John Illingsworth.

About the same view of Collaroy-Narrabeen Beach 2023 with a $25m seawall in place designed to protect 36% of the beach and 49 properties.

Storm damage to shore platforms and cliffs is rarely mentioned in the media, but it certainly occurs. A study of rock falls from the cliffs between Long Reef and Barrenjoey found that the average volume of rock falls was six times the previous estimate and that rock falls were a natural hazard that Local Government should be concerned about. Nearly all falls occur during big storms and/or heavy rainfall. Those same conditions severally impact the biological communities on shore platforms. So too does human predation of edible shellfish which can be quite heavy even in legally protected areas such as Long Reef, but that can be addressed through education. It can be quite depressing to visit Long Reef after a big storm or a sunny weekend with lots of visitors, the place can look like a disaster zone only lacking refugees and blue tarpaulins.

Large rockfall from 65m high Avalon Head in 2017. Photo taken in 2021 by John Illingsworth.

Three months later normality seems to have returned and although the details of species distribution are different, we are always amazed at the resilience of the system. That's fine for this storm, and the next storm, but what about the cumulative impact of sequential storms over decades as sea level gets higher and we experience other events like heatwaves, or the arrival of an exotic species? This question is impossible to answer but the one thing we can be sure of is that the different biological zones and patches on the platform will shift. The platforms are confined by cliffs, the inter-tidal zone must become narrower and species populations must decline with a likely increase in extinctions.

To some extent we can offset future species losses by thoughtful design of the inter-tidal built environment such as sea walls and wharfs. Work in Sydney Harbour illustrates this concept. Most of the harbour is not an open ocean coastline and conditions of water quality and waves are different but the original rocky shores did carry functioning inter-tidal communities. Beginning in the 1840s more and more sea walls were built to formalize the harbour shores, and to close-off tidal inlets (salt marsh and mangrove communities) and 'reclaim' land. Most sea walls were constructed of sandstone blocks with a vertical face, in more recent times concrete, steel and even plastics have been used and these materials provide little opportunity for species colonization. Sea walls like these offer a reduced surface area with little surface textural variety and are only slowly colonized by a limited number of algae and invertebrates in comparison to natural rock. Two simple design changes can make a big difference. Old walls can be retrofitted with moulded concrete panels that provide better physical niches as achieved by 'Living Sea Walls'. Monitoring these structures has shown a 30-40% increase in species diversity and a significant shift in ecosystem services such as water filtering, and fish food resources.

A 'living sea wall' attached to an older nearly barren concrete wall in Sydney Harbour. They work, and examples have now been installed overseas.

Another approach can be seen in the landscaping of Barangaroo Headland on what used to be extensive industrial wharves. The sea wall concept was entirely rethought and rather than present a vertical face to the water the edges were constructed of thousands of large sandstone blocks placed as uneven steps down into the water, Not only is this an attractive constructed landscape but the large increase in sub-horizontal surface and the numerous crevices and uneven block surfaces has greatly enhanced colonization by marine organisms. The blocks allow human access to the water's edge even though they can be slippery, kids (and more agile adults) love them!

This simple design shift created a surprising constructed shore platform, that will be ecologically self-sustaining and satisfying to both the marine colonizers and human visitors.

A constructed sea wall on Barangaroo Headland enriches both the inter-tidal habitat and visitor's experiences.

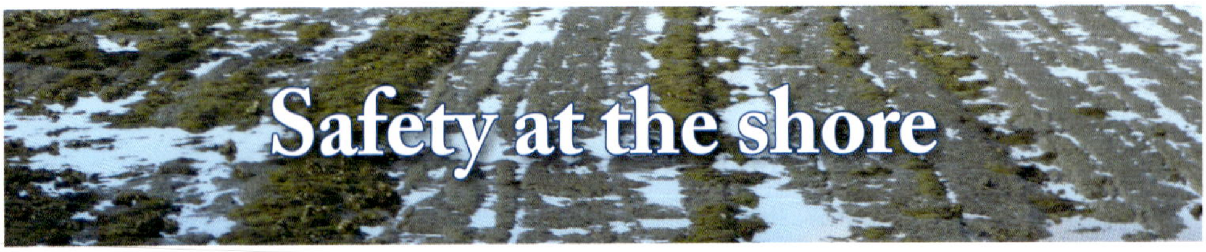

Safety at the shore

Coming home unharmed after a day on the coast is highly desirable, so we have listed a number of safety tips that you should be aware of. The most important of these is simply to use your common sense. We all have it, but it does not always engage☺.

The hazards you may meet and the precautions you should take

General

- Know what the weather is doing, and be aware of the tide times.
- Take note of any safety signs you see.
- Be aware of the location of rescue equipment if it is present.
- Supervise children.
- Carry a first aid kit appropriate to the size of your group.
- Know the emergency phone numbers, 000 is the best but if you only have a mobile that is out of range you can often still get through by dialling 112.
- Know where you are in case you do need to summon assistance. Remember the name of the street you parked in and have a clear description of your location.
- Watch your valuables. Theft from cars and from under towels on the beach is all too common.

Big waves, rough water, and rip-currents

- We trust that you can swim and that you are not venturing into risky locations alone.
- Observe the scene for five or ten minutes before you venture far and always watch the waves and tide.
- If swimming from the beach do it at a safe place, ideally between the red and yellow flags, and know how to recognise a rip-current.
- If snorkelling or SCUBA diving near the rocks you should have undertaken appropriate training and always dive with a buddy. A wet suit may be desirable.
- If you are unsure of conditions, ask a local or back off, and come back another day.
- If you are working near the surf it is wise to wear a life-jacket, in declared areas of NSW rock fishers are now legally required to wear a life jacket.
- Select an escape route before venturing into any tricky situation.
- If you do accidentally end up in the water swim away from the rocks and look for a safe place to come ashore.
- **The bottom line - never turn your back on the sea.**

Sunburn and dehydration

- Avoid exposure during the hottest part of the day.
- Wear appropriate light clothing, sunglasses, long sleeves, and a shady hat. So you look dumb, but you will go home smiling!
- Apply plenty of water resistant sunscreen with a sun protection factor (SPF) of at least 30+.
- Remember you get a double dose of UV reflecting from the water and the sand, and a hot wind can make it even more uncomfortable.
- Carry sufficient fresh water to keep you hydrated.

Trip and slip hazards on the uneven surface

- Never run on the rocks but walk carefully. Wet rock surfaces, covered in a film of algae can be extremely slippery.
- Bare feet are not wise as they are likely to be cut.
- Cleated boots such as rock fishermen sometimes use can damagc the organisms you have come to see.
- Thongs are just too slippery so you need to find a pair of shoes with non-slip soles (test them on a smooth wet surface) and made of a material that will stand long immersion in salt water. Each to his own, but we find that old-fashioned tennis shoes such as Dunlop Volleys do a good job. Sorry, they are not a fashion item (unless you are a 14 year old male).
- Don't use waders or rubber boots. They can be slippery, they may tempt you into dangerous water and they limit your ability to recover or swim if you do enter deep water.

Rock falls

- Avoid working under cliffs during or after heavy rain.
- Watch for early signs of falls such as frequent small stones tumbling down, or evidence of recent big falls.

Cuts and abrasions

- Easy to get on oyster shells, barnacles and rough rocks.
- Good footwear and watching where you place your feet and hands are the answers.

Venomous and nasty beasts

- Yes, they exist but with a little knowledge you are not likely to come into contact with them.
- Be very wary of placing your fingers under loose rocks, into crevices, or dense seaweed where you can't see what may be lurking there.
- Don't pick up or touch organisms you know to be dangerous (bluebottles, fire-worms, blue-ringed octopus, and cone shells in more tropical waters.
- Handle empty shells and soft drink cans in the water with care as they might contain an octopus.

Important first aid

Treating Sunburn

Sunburn is caused by ultra-violet radiation and can occur even on cool and cloudy days. UV radiation can penetrate deep into the skin, damaging the cells, and setting you up for skin cancers later in life, so take it seriously. Prevention is best, but if it does happen, as soon as you are aware of the damage avoid further exposure by covering up and moving indoors, or at the very least apply more sunscreen.

- Drink plenty of water, but avoid very cold drinks.
- Gently apply soft pads dipped in cold water or cold black tea to the burned surfaces.
- Take a cool shower or bath.
- Seek advice from a pharmacist or doctor about pain relief or blistering

Bluebottle stings

If you tangle with the stinging tentacles of a bluebottle you will instantly know it as they are quite painful, and even ‘dead’ ones on the beach can sting. Old remedies such as rubbing the stinging cells from the skin with a handful of sand or a towel actually make the pain worse.

The recommended first aid is to gently pick off attached tentacles with tweezers or a gloved hand, and to wash or immerse the site with plenty of hot water, but not hotter than 45^{0}C (that is as hot as you can tolerate without burning). Ice packs may relieve some pain but hot water has been found to be superior (don’t forget your thermos). The application of vinegar, seawater, or alcohol is not advised.

Serious stings to the face and neck, or in cases where breathing or heart rate is impaired and shock is evident require an ambulance. **Expired air resuscitation (mouth-to-mouth) and cardiac massage (formerly called CPR) may be needed if breathing or heart beat fails. Remember the ratios? Two full breaths are required for every 30 chest compressions applied at a rate of 90 compressions per minute.**

Jellyfish

Box and Irukandji jellyfish are only likely to be encountered in tropical waters during the wet season and the best approach is to avoid entering the water during the danger period. Box jellyfish stings can be fatal and urgent medical assistance is essential. Jimbles seem to only like very sheltered bays in temperate zones and don't pack the wallop of their northern cousins.

Flood the affected area with vinegar (**not** methylated spirit or alcohol). Then remove tentacles gently with a gloved hand or tweezers. Medical assistance should be summoned urgently. Expired air resuscitation may be needed if the victim's breathing fails.

Blue-ringed octopus

A bite from the blue-ringed octopus is not painful and may not even be noticed. You may feel a slight prick and a small drop of blood may appear at the bite site. Within ten minutes symptoms will include general weakness, numbness around the face, vomiting, tightness in the chest, difficulty in breathing. Death may occur in as little as thirty minutes. **Summon medical assistance urgently by calling 000** or if your mobile is out of range try 112.

There is no antivenom available and pressure-immobilisation is the recommended first aid.

If the bite area can be identified apply a firm crepe bandage, and a second bandage to immobilise the limb. Begin at the toes or fingers and work up firmly toward the heart. The victim should rest and be observed closely. Recovery is possible but expired air resuscitation and cardiac massage may be needed to sustain life until professional help arrives.

For more information concerning marine stings contact the Australian Venom Research Unit 24-hour help line on 03 9483 8204 or check their web site.

Glossary

Abdomen – the rear segment of a crustacean body. The edible part of a lobster or prawn!
Aerobic and anaerobic – well aerated or totally lacking in oxygen.
Albedo - the proportion of solar radiation that is reflected into space from a surface such as the Earth, the ocean, or clouds.
Alga (pl.algae) – a plant without roots, stem or leaves that can photosynthesise. Includes almost all seaweeds.
Amino acids – organic compounds that are the building blocks of protein and life.
Anthropocene - the present geological age when human activity has been the dominant influence on climate and the environment. A precise starting date has yet to be formalised.
Ascidian – sea squirts or tunicates belonging to the class Ascidiacea in the phylum Chordata.

Barren grounds – a bare rocky sea floor environment that support almost no life forms, particularly those created by heavy grazing by sea-urchins.
BCE – time before the common era, formerly BC before Christ.
Biodiversity – the diversity of all biological materials at any level of organization, genes, species or communities.
Bioluminescence – the biochemical emission of light by living organisms such as algae or glow worms.
Biomass – total mass (weight) of plants or animals in a given place or environment.
Bivalve – a mollusc with two hinged shell valves.
Budding – reproductive process that creates new individual plants or animals by detaching small buds from the parent. All individuals are genetically identical (clones).

Calcareous – composed of calcium carbonate $CaCO_3$. Two mineral forms occur, aragonite and calcite.
Cambrian – geological period in which organisms with hard parts proliferated and fossils became abundant. Extended from 542 to 488 million years ago.
Camouflage – mechanisms for making an organism become less visible against its background.
Canopy – the upper layer of leaves or fronds in a plant community.
Carapace – the outer 'shell' or exoskeleton of a crustacean.
Cerata – soft finger like projections on the body of sea slugs.
Chitin – a horny organic substance with a composition similar to hair that forms the exoskeleton of crustacea, and the operculum of some gastropods.
Chloroplasts – organelles containing chlorophyll within plant cells that are the site of photosynthesis.
Chromoplasts – organelles containing pigments other than chlorophyll within plant cells that are the site of photosynthesis.
Class – major subdivision of a phylum.
Concretion – a spherical body of cemented sediment enclosed within a sedimentary rock. Often formed by accumulation of a cementing agent around a fossil.
Coriolis 'force' – an imaginary force used to explain the direction in which objects move relative to the rotating earth. Applies to currents and winds but not to waves.

Desiccate – to dry out.
Dispersal – movement of organisms to another place.
DNA - deoxyribonucleic acid, the molecule carrying the genetic code in almost all life forms.
Dyke – near vertical sheet of igneous rock that has intruded other strata.

Ecology – study of the web of life.
Eddy - a circular movement of water, a vortex.
El Niño – major change in ocean surface currents that affect the weather. An increased flow of warm water that brings high rainfall to the deserts on the western American coast and an increase in cool water in eastern Australia that is associated with drought. Also referred to as ENSO. El Niño/Southern Oscillation. La Niña is a reversal of the El Niño pattern.
Encrusting – growing over the substrate like a sheet.
Epiphytes – an organism that grows on another but does not get any sustenance from the host.
Exoskeleton – external protective cover or skeletal structure as in the crustaceans.
Eulittoral – mid region of the shore that is exposed and submerged during each tide cycle.

Family – subdivision of an order, generally containing a number of genera.
Flagella – long whip-like filaments in many organisms that enable movement or create water currents as in the filter-feeders.
Flotsam – material that has floated on the water, originally related to ship's wreckage.

Gaia hypothesis - the hypothesis that proposes that living organisms interact with their surrounds to form a self-regulating system that maintains the conditions for life on Earth.
Gamete – germ cells that unite to form a new individual, as in egg and sperm.
Genus – major subdivision of a family in which a number of species are placed because they are very similar to one another.
Ghost nets – lengths of plastic fishing net that have been cut adrift. They continue to trap fish, dolphins and turtles until they wash up on the shore.
Globose – globe shaped.
Gyre - a large system of circulating ocean currents occupying up to half of any ocean. Driven by winds and modified by Coriolis and friction.

Habitat – the place where an organism finds all the necessities for life, not to be confused with niche
Hermaphrodite – an organism that has both male and female reproductive organs.
Homing – ability and habit of an organism to return to the same place.
Hypersaline – having much more salt in solution than normal seawater.

Igneous rocks – formed from a rock melt in the earth's crust, may be volcanic rock like basalt, or pumice, or may be plutonic rock like gabbro or granite.

Jetsam – goods thrown overboard to lighten a ship in distress.
Joint planes – regular pattern of fractures in rocks that are lines of weakness along which weathering will advance more quickly.

Keystone - an organism that holds the system together. Loss of a keystone species can trigger extreme change in an environment.

Larva (pl. larvae) – juvenile animal that has a different body form from the adult.
Littoral – the intertidal zone.
Littorinids – members of the family Littorinidae, molluscs that dominate the upper littoral.

Mantle – the soft tissue of molluscs that lies next to the shell and can often extend to partly cover the outside surface of the shell. This organ secretes the shell with the inner layer being either like porcelain or mother-of-pearl (nacre).
Metamorphic rocks – any rock that has been subject to significant change as a result of heat or pressure. Sandstone for example will convert to quartzite, limestone to marble, and mudstone to slate.

Metazoa – multi-celled animals.
Motile – capable of moving.
Moult – to cast off the skin or exoskeleton and thus enable an animal to grow. Normal in the Crustacea, also seen in some algae.
Mucus – slime produced by an animal.

Nano – microscopic particles smaller than one billionth of a metre.
Neuron – nerve cells.
Niche – the ecological role of an organism in a community, not part of the habitat.
Notochord – a rod-like structure found in the body of protochordates that is similar to that found in the embryos of vertebrates.
Nurdles – industrial pellets of plastic two to five millimetres in diameter used as a feedstock in plastics manufacture.

Ocean acidification – a poor term for the observed decrease in average ocean water pH from 8.15 to 8.05 between 1950 and 2020. Driven by increased carbon dioxide emissions into the atmosphere.
Omnivore – an animal that eats both plants and animals.
Operculum – the disc like structure that closes the aperture of barnacles, polychaete tubes and many gastropods.
Oral and aboral – the surface on which the mouth or anus is found.
Order – major subdivision of a class.

Palaeontologist – a person who studies fossils.
Patch dynamics – processes that cause patterns in the distribution of organisms through their interaction with one another.
Pelagic – animals that live at or near the ocean surface.
Permian – period of geologic time that extended from about 299 to 251 million years ago.
pH – measure of acidity or alkalinity. pH 7.0 is neutral and the sea is usually mildly alkaline at pH 8.0
Pheromones – scent-like compounds secreted by organisms as a means of chemical communication.
Photosynthesis – process that uses light energy to synthesise complex molecules from carbon dioxide and water.
Phylum – organisms with the same general body plan.
Pixel – the smallest graphic element in a computer screen, pixellation colour patterns employing this concept that make very effective camouflage.
Plankton – plants and animals that have limited ability to move by themselves and that are transported by currents.
Pleistocene – an epoch of geologic time from about 1.8 million years ago to 12,000 years ago. Followed by the Holocene or Recent.
Proboscis – a non-rigid snout-like feeding organ.
Phytoplankton – plant plankton.

Radula – a tongue like organ in the molluscs that is lined with many hard 'radula teeth' that rasp off food.
Reef - a ridge or shoal of rock, coral, or shellfish at or near the surface of an ocean or bay.
Rhizoids – root-like filament of a holdfast.
Rip-current – a dangerous rapid flow of water away from the beach, best identified by the absence of breaking waves and apparently calm water.
Rock stacks – isolated columns, small cliffed islands, or piles of rubble that have been isolated from a cliff.
Rock structure – three dimension patterns of the planes of weakness in a rock mass defined by bedding, joints, faults and folds.

Saltmarsh - ecosystems vegetated by herbs, grasses or low shrubs bordering saline waterbodies on the coast they occur immediately above mangroves and below terrestrial vegetation.
Sedimentary rock – rocks composed of cemented sediments, for example; conglomerate, sandstone and shale.
Sessile – organisms that are attached to the sea bed.
Seta (pl. setae) – bristles extending from the body of annelid worms.
Siphons – muscular tubes in animals that draw in water and food and expel wastes.
Spat – oyster spawn or young oysters.
Species – the basic kind of biological category, a single kind of organism.
Spicules – needle like crystals that form the skeleton of sponges.
Spore – single celled product of reproduction that can develop into a new individual organism.
Stipe – the unbranched 'stem' of an algae that connects the holdfast to the frond.
Sublittoral – lower region of the shore only exposed by retreating waves at low tide.
Substrate – the rock or sediment to which sessile organisms are attached.
Succession – the sequence of replacement of one organism or assemblage by another .
Supralittoral – highest region of the shore wetted by wave splash at high tide.
Swell – background pattern of waves that are decayed remnants of distant storm waves.

Temperate – mid-latitude region between the poles and the tropics.
Terrestrial – living on the land.
Test – skeletal parts ('shell') of echinoderms or foraminifera.
Thorax – segment of the body between the head and the abdomen in crustaceans. The thorax usually carries the legs.
Triassic – period of geologic time after the Permian that extended from about 251 to 200 million years ago.
Trophic – related to food and feeding.
Tsunami – major ocean wave generated by an earthquake, volcanic event or sediment slumping. The wave has a long period and can reach extreme heights at the coast. Incorrectly called a tidal wave.

Umbo – apex or point of a bivalve shell just above the hinge.
Upwelling – where surface waters are replaced by cold, nutrient-rich water that 'wells up' from below. Occurs most often along the coast when winds blow parallel to the shore.

Vascular system – in echinoderms a complex of ducts within the body filled with water that acts as a hydraulic network that enables movement, food gathering and attachment.
Vesicles – a bladder like structure in some algae. May be filled with gas or water and helps the algae to float, or resist desiccation when exposed to the air.

Water layer weathering – rock weathering on a shore platform driven by alternate wetting and drying and salt crystallization.
Wrack – seaweed and other debris that has been cast ashore.

Zonation – pattern of distribution of different organisms that reflects different levels of wetting by tides and waves.
Zooid – the basic part of a bryozoan, bluebottle, or ascidian colony. Different zooids are specialised for different roles.
Zooplankton – animal plankton.
Zoospores – algae spore that can move by means of flagella.
Zygotes – the initial cell of a new individual created by the union of two gametes.

Resources

This is a short selection of texts that may help answer some of your questions. You will also find a huge amount of useful material on the web, and it is often easier to dig there first.

Andrew, N. (ed.) 1999. *Under southern seas. The ecology of Australia's rocky reefs*. Univ. of NSW Press, 238 p.

Attenbrow, V. 2010. *Sydney's Aboriginal past. Investigating the archaeological and historical records.* UNSW Press, 2cnd Edn. 226 p.

Bennett, I. 1992. *Australian seashores: adapted from W.J. Dakin's Australian seashores.* Collins/Angus and Robertson, 266 p.

Brusca, R.C. and Brusca G.J. 2003. *Invertebrates.* Sinauer Assoc. Inc., Publishers, Mass., 2cnd Edn. 936 p.

Byrne M., and O'Hara Tim. Eds. 2017. *Australian Echinoderms: Biology, ecology and evolution*, CSIRO Publishing 624p

Carefoot, T., with Simpson, R.D. 1977. *Seashore ecology. Adapted for Australian conditions.* Univ. of Qld. Press, 278 p.

Carson S. and Morris R. 2017.*Collins Field guide to the New Zealand Seashore*. HarperCollins NZ. 426.

Chapman, A.D. 2009. *Numbers of Living Species in Australia and the World.* Australian Biodiversity Information Services, 2cnd Edn. Also on the internet.

Coleman, N. 1977. *A field guide to Australian marine life.* Rigby, 223 p.

Coleman, N. 1981. *What shell is that?* Landsdowne Press, 298 p.

Dakin, W.J. (with Isobel Bennett) various editions 1952-1987. *Australian seashores.* Angus and Robertson, varies 372 to 410 p.

Davey, K. 2004. *A photographic guide to seashore life of Australia*. New Holland, 144 p.

Edgar, G.J. 2001. *Australian marine habitats in temperate waters.* New Holland, 224 p.

Edgar, G.J. 2008. *Australian marine life: the plants and animals of temperate waters*. New Holland, 2cnd Edn. 624 p.

Fuhrer, B., Christianson I.G., Clayton M.N. and Allender B.M. 1988. *Seaweeds of Australia.* Reed Books, 112 p.

Jones, D.S. and Morgan, G.J. 2002. *A field guide to Crustaceans of Australian waters.* New Holland Publ. P/L, 2cnd Edn. 224 p

Jones D., and Morgan G. 2007. *A field guide to crustaceans of Australian waters*. Reed New Holland, 224 p

Marsh L.M. and Fromont J. 2020. *Field guide to shallow water seastars of Australia.* Western Australian Museum. 544 pages

Miskelly, A. 2002. *Sea-urchins of Australia and the Indo-Pacific.* Capricornia Publications, 179 p.

Morton, J.E., Scientific editor B.W. Hayward. 2004. *Seashore ecology of New Zealand and the Pacific*. Bateman, 504 p.

Morcombe, M. 2003. *Field guide to Australian birds.* Steve Parish Publishing, 448 p.

Pepperell, J. 2001. *Fish tales. The mystery of the snapper bump and other stories from Australian waters.* Random House Aust., 206 p.

Porter, C.M., Westcott, G.C. and Quinn, G.P. 2010. *Life on the rocky shores of south-eastern Australia.* Vict. National Parks Assoc. 128 p.

Porter, C., Mathews T.G., Bellgrove A. and Westcott G. 2023. *Field guide to the seashores of south-eastern Australia.* CSIRO Publishing 228 p

Short, A.D. and Woodroffe, C.D. 2009. *The coast of Australia.* Cambridge Univ. Press, 304 p.

Underwood, A.J. and Chapman, M.G. 1993. *Coastal marine ecology of temperate Australia.* NSW Univ. Press, 341 p.

Underwood, A.J. and Chapman, M.G. (eds.) 1994. *Seashores. A beachcombers guide.* NSW Univ. Press, 117 p.

Wilson, B. 2002. *A handbook to Australian seashells on seashores east to west and north to south.* New Holland Publishers, 185 p.

Wilson, B.R. and Gillett, K. 1979. *A field guide to Australian shells. Prosobranch gastropods.* AH and AW Reed, 287 p.

Woodroffe, C. D. 2002. *Coasts: Form, process and evolution.* Cambridge University Press, 623 p.

Web Resources

A photoguide to species. Morton, J. K. (2020). Intertidal rocky shore life: An identification guide for the Central and Hunter Coasts of NSW, Australia. Avondale University College. Cooranbong. https://research.avondale.edu.au/sci_math_books/5/

A virtual tour: Living sea walls Sydney Harbour. 4 min video. https://www.youtube.com/watch?v=CA056FuQIOI

Australian Shellfish Reef Restoration Network (SRRN) https://www.shellfishrestoration.org.au/

Australian Venom Research Unit https://biomedicalsciences.unimelb.edu.au/departments/department-of-biochemistry-and-pharmacology/engage/avru/advice-and-resources/first-aid#two

Coastal Environment Centre https://www.northernbeaches.nsw.gov.au/environment/coastal-environment-centre

Living sea walls https://www.livingseawalls.com.au/

Marine Education Society of Australia with good images and stories. http://www.mesa.edu.au/

New Zealand molluscs https://collections.tepapa.govt.nz/topic/965

Pittwater Pathways LONG REEF: special by any measure. 35 min video. https://www.youtube.com/watch?v=w7PsLBkxWVs

Pittwater Pathways WRITTEN IN STONE: a 245 million year old mass breeding event? 32 minute video.

Proper names for marine fauna, site maintained by CSIRO http://www.marine.csiro.au/caab/

Seashells, an Australian Museum resource compiled and managed by Des Beechey. http://seashellsofnsw.org.au/

Sea slug forum http://www.seaslugforum.net/ hosted by the Australian Museum.

Sea surface temperature http://www.bom.gov.au/marine/sst.shtml

St Johns first aid - latest first aid fact sheets https://stjohnnsw.com.au/first-aid-guides

Tide times for NSW http://www.bom.gov.au/oceanography/tides/MAPS/nsw.shtml

Index